Use of ICTs in Agriculture Extension

Use of ICTs in Agriculture Extension

DR. NIKULSINH M. CHAUHAN

BIOTECH
2012
BIOTECH BOOKS

ISBN 978-81-7622-261-7

Published by : **BIOTECH BOOKS**
4762-63/23, Ansari Road, Darya Ganj,
New Delhi - 110 002
Phone: +91-011-23262132
E-mail: biotechbooks@yahoo.co.in

Laser Typesetting : **Sushil Kumar**

Printed at : **Chawla Offset Printers**
Delhi - 110 052

PRINTED IN INDIA

Dedicated to my beloved
Late Brother Dilipsinh Chauhan

Acknowledgement

This memorable occasion provides me with a unique opportunity to express my profound sense of gratitude and reverence to my Respected research Guide **Dr. Narendrasinh B. Chauhan**, Ph. D., Professor & head, Department of Extension Education, B.A. College of Agriculture, Anand Agriculture University, Anand for his talented guidance, persistent and inspiring encouragement, constructive suggestions and keen interest throughout the present study, construction of my brilliant academic records, development of higher vision and also for preparation of this manuscript as well for preparation and submission of all the four Satellite research Project cum PGR Assignments as a whole. I am also thankful to all of his family members for giving me a moral support during the course of my study.

I am deeply thankful to **Dr. Surya Gunjal**, Director, the School of Agricultural Sciences, Yashwantrao Chavan Maharashtra Open University, Nashik for giving me an opportunity to be a part of postgraduate studies and rendering me continuous advice, help, inspiration, motivation and encouragement during the period of my Ph.D. study.

I would be unworthy if I fail to express my cordial thanks to **Dr. Somnath Suryawanshi**, PGR Coordinator, PGR Programme, YCMOU, Nashik for their valuable help, Kind cooperation at each and every stage of this Ph .D. study mission, giving continuous and constructive suggestions thought out study and providing required tenderness whenever, I needed during the whole journey of my higher study.

I am also thankful to all the Teacher Counselors, experts and Evaluation Committee Members of YCMOU, Faculty members of the school of agricultural Sciences, YCMOU, Raosaheb Patil, Training organizer, KVK YCMOU, Nashik, Guest House staff and all Colleagues of PG Study, especially Professor Y. S. Patil, Professor Lahoti and Rajabhan Kakde.

I treat it to be the great privilege to record my heartiest thanks to Yashwantrao Chavan Maharashtra Open University for giving me prospect to be a Ph. D. holder otherwise my aspire to be a Doctorate would remain a dream only.

My humble whole hearted thanks to Dr. A. M. Shaikh, Principal and Dean, B.A. College of agriculture, AAU, Anand, all faculty members of the Department of Extension Education, B. A. College of Agriculture, Anand, Dr. P. N. Upadhyay, Principal and dean, N. M. College of Agriculture, NAU, Navsarai, Dr. G. B. Valand, Dr. G. N. Patel, Dr. C. K. Dixit, Dr. J. G. Patel, Mr. Kantibhai Patel, Dr. Maheshbhai R. Patel and Mr. Vijaybhai Darji all of AAU, Anand Campus for their valuable support, constant inspiration, motivation and encouragement to me during my Ph.D. study.

My humble whole hearted thanks and extreme gratitude to honourable **Mr. S. N. Batliwala**, The Executive Secretary and Chief Accountant, Sir Dorabaji tata trust, Bombay, **Mrs. R. M. Marawala** and **Mr. Axaykumar Ramanuj** for providing me an Education Grant in the form of Scholarship to complete this study. I would be unable to complete this Doctorate Degree without their financial support. I heartily acknowledge their valuable help for my academic and Research work. They have given wholehearted support during my study period.

I shall remain extremely thankful to Mr. Dilipsinh Sisodiya, Shri. K. S. Jadav, Ashok Gohel, Rajiv Ghosh, Sunil Parmar, Hiren Prajapati, Shambhu Rathod and all of my P. G. Colleagues of B.A.C.A. AAU, Anand, for their ever willing help from the inauguration of the fire stone to completion of the Ph. D. building. I am also obliged and record my grateful thanks to Mrs. Diptiba Chauhan and Chiranjivi Kuldeepsinh Chauhan for their moral support during the entire period of my postgraduate study.

I owe a special debt to all training organizers of KVKs in India for giving me valuable, timely and accurate information for this investigation. I hope the findings of the study will be useful to them in managing their KVKs in a better way.

The excellent and untiring effort made by Shri. Rajesh Makwana by providing me the technical support in typing, collection of the photographs, references and required information from internet and computer are duly acknowledged.

I express my sincere thanks to Dr. B. V. Padhiyar, Associate Professor, Aspee College of Horticulture and Forestry, NAU, Navsari for his moral and technical support during my Ph.D. study.

I take this opportunity to express my deep sense of gratitude and profound regards to my beloved mother Gangaswarup **Gajraba**, Elder brother **Mr. Gunvantsinh Chauhan**, Mama **Shri Khushalsinh Virpura**, Retd. Deputy Collector, **Mr. Balvantsinh Sisodiya** my father in low and his family members for their affection and moral support to build up my educational career. Without the motivational and warm beams from them the construction work of Doctorate Degree would remain a daydream.

Above all, I have no words to express my gratefulness to my better half **Mrs. Rajeshrikunverba**, my son **Preetpal** and Daughter **Shubhangini** for their patience, inspiration and decent support and allowing me to accomplish my work by helping me through performing some of my domestic and social responsibilities during the period of my study, without which this could not have been a reality. I am highly appreciative of the personal sacrifices made by **my kids and my wife** once again. My special thanks to all of them who had directly and indirectly helped me on thorny path for the making this dream possible and come true to be a Doctorate.

Place: Anand **(Nikulsinh M. Chauhan)**

PREFACE

In the Indian perspective, agriculture should develop in to an effective instrument for creating income, employment and food security. These triple goals could be achieved if we generate and disseminate the agricultural technologies that are relevant to the end user. However, neither technology generation, nor technology transfer take place in vacuum. Both these processes are dependent upon each other. Similarly, both the processes are influenced by and do influence social, cultural, economic, political and the like aspects in the society. Our country has come a long way after independence, but still faces the problems or poverty, hunger and malnutrition. The globalization, liberalization and privatization has posed new challenges and created new problems in certain aspects of agriculture. Therefore, time has to examine our achievements and failures in agriculture in the context of this changing environment. Technology development and technology transfer are the two crucial processes in agricultural development and socio-economic enlistment of the farming community. So, there is need to think over the various dimensions of technology development and technology transfer in agriculture and allied sectors.

The KVK is the front line transfer of technology centers of ICAR. The KVKs are mainly made to cater the needs of vocational trainings to farmers, farm women, Rural youths and extension personnel's to increase their agricultural production, increasing the income and as a result of this to raised their standard of living. The KVK is also assigned the task of technology generation, technology assessment, technology refinement and technology dissemination in the field of agricultural extension. The KVK is the best jewelers of ICAR at grass root level. KVKs are working on the base of two main pillars of extension education, **"Seeing is Believing" and "Learning by Doing".**

Salient Features for KVK Fortification as perceived by myself are each and every SMS should have at least one independent /identical project to create their own identity and KVK as a whole. At least one success story in their concerned field of work within year or two. Concentrate all TOT activities based on two main Pillars of the Extension Education. Develop as a Real Information Hub of the District. To create Farmers friendly Environment and Maintain each and every Demonstration units on Scientific Line. **Farmers Friendly, Farmers Centric & Farmer Led Extension approach** should be adopted and develop **F-R-E-M** Link based extension. Rapid TOT in Local Language through active participation of the Local leader, Resource persons and Rural youth in TOT process. Extension with input supply. Intensive efforts for injecting NCF Approach. Preservation of Land, Water and Environment for Sustainable Agriculture Development. To develop Website Services, FAQs-Database, ICQAP & EQAP Agricultural products, Seed supply, Year round, Vegetable Nursery, Bioagents/Vermiculture/ Biocompost, Integrated Approaches in all Management of Agricultural Extension Management .Precision Farming ,**Effective Recycling of Farm Waste- A Call of new Millennium**. This book covers all of the gentle ideas of effective TOT in chewable manner. It is real manual based on field extension experiences. This book can explain the use of ICTs in agricultural extension and clear cut guideline for TOT with the help of modern ICT Tools.

I am working in KVKs since last sixteen years. Under my leadership KVK Vyara, Dist- Tapi is really become an Information Hub for farmers. The long ways of successes are presented here. I congratulate to all of my SMSs for building of the road up to here with hard working. I am very much thankful to NAU Authority and all GOs, NGOs, Cooperative sectors, and all line departments related to agriculture and rural development to cater their help, suggestions, warmth, guidance and also bitter comments of our works. Based on my vast experience in the field of Transfer of Agricultural Technology I have written this book. I hope this book" Use of ICTs in Agriculture Extension" will be a boon for all concerned with TOT work.

I have tried my level best to make this memorandum up to the mark. However, any error in the matter will be a part of my responsibility. Your constructive suggestions in the matter are highly solicited and appreciated, too.

(N. M. Chauhan)
Programme Coordinator
KVK, NAU, Vyara

Contents

1 Opinion and Expectations of the Farmers Regarding the Use of Internet for Agricultural Extension

EXPECTATION OF THE FARMERS TOWARDS ICT IN AGRICULTURE AT VILLAGE LEVEL

ABSTRACT

The scientific information and its effective communication by different scientific departments of agriculture, college of agriculture and other developmental organizations are considered to be the most important ingredients of agricultural and rural developmental strategy in Agrarian County like India. The communication requires a series of interrelated and transmissible line system to generate and diffuse new information and technology innovations. Among all the available means for speedy communication, use of Internet has been emerging as powerful and versatile media. The Government of India has made intensive efforts to meet the future Challenges in various developmental fields, T & V system is one of them. The ratio of VLW and farmers is now 1: 1000, thus it is quiet difficult to reach the unreached by this channel. This gap remains a challenge for the extension system even today. To reach over 110 million farmers, spread over 500 district and over 6000 blocks is an up hill task. The diversity of agro ecological situations adds to this challenge, further farmer's needs are much more diversified and the knowledge required to address them is beyond the capacity of the grass root level extension functionaries. Today it is possible to find a solution to this situation by using the potential of satellite based Internet technologies to meet the location specific information needs of the farmers. Information and communication networks are expanding very fast. The number of satellite based Internet connections in India has crossed the two million mark and the number of Telephone connections is over 320 million.

It is also true that Internet is not absolute solution for effective extension of rural and agricultural development but it does bring new information resources and can open up new communication channels for rural communication and agricultural organizations. It offers a means for bridging the gaps between developmental professionals, rural people and agricultural producers through the initiation of interaction and dialogue.

There has been a rapid increase in the use of the internet in developing countries, with regard to rural and agriculture development and the internal, extension organization that support the betterment of rural populations and improvements in agriculture have important roles to play for the benefits of improved communication and access to information. Thus to take a maximum benefits of the system this study was selected to know the expectations of the farmers towards Community Internet Centre at village level. By knowing this we could be able to establish the rural Internet centers on need based approach. The study was undertaken in purposively selected Anand taluka of the Anand district of Gujarat in a random sample of total 100 farmers from purposively selected four villages. The data were collected by personal contact method. The tool used for study was personal interview schedule.

1. INTRODUCTION

India, the world's largest democracy, focal point of competition between rival empires local and foreign for more than four thousand years; the world's most populous, after the people's republic of China is predominantly a nation of various religions, classes, casts, languages, cultures and societies. Agriculture continues to be the major occupation and way of life for more than half of its population even today. Sustainable prosperity of this class, the farmers, and the landless agricultural laborers holds the key for improving the overall human resource development scenario and ultimately the wealth of the nation too.

The economy of our nation is based on agriculture and allied agro based industries, animal husbandry and all other agriculture based businesses. Thus overall development of our country is not possible by ignoring the agriculture and farmers in the process of development. Present time is an age of information communication technology, thus perfect, quick, timely, required and understandable information in the field of agriculture has been considered as indispensable input.

Indian agriculture had been on traditional lines till the first waves of Green Revolution in the late sixties. The Green Revolution gave a sudden boost to the production and productivity of major cereals in the assumed irrigated areas. The major segments of our population are being lived in rural areas. Hence, quick dissemination of technology information from the agricultural research system to the farmers in the field and reporting of farmer's feedback to the research system are the critical inputs in transfer of agriculture technology. To disseminate the required and recent agricultural information to the farmers in scattered villages at the variegated geographical situation in India is very difficult task. Though, the Training and Visit system is under implementation and the VLW-farmer; ratio is about 1: 1000, it is quiet difficult to reach the unreached by this channel.

Looking to the importance of the agriculture in national economy, high level efforts are needed to stand in open global markets. For this, quality production, value addition, integrated approaches and all recent innovative technology dissemination are required to be thought-out. India had made a remarkable progress in food grain production, even though to take the challenge to head our burgeoning population in next decade is a risky task. It is recognized that there is a wide gap between the present level of yields and the achievable potential both in irrigation and rainfed areas. Lack of effective agriculture extension agency for transfer of technology to the farmer's fields is one of the major constraints impeding the effort for increasing agricultural production.

The present rate of agricultural production could be doubled if the available technologies are brought to bear with production process and programmes. Focusing more and more on transferring our new technologies away from the confines of laboratories and research institutes to the farmers can make the situation more result and work-oriented. The present level of production can be increased from 50.00 per cent to 200 per cent, only by good extension efforts for promoting precision in farming and farm management.

Transfer of technology to the level of farmers is not a one time exercise because new farm technology is being constantly evolved. A continuous flow of technologies in an appropriate manner is vital to provide quick benefit of this development to the farmers. There has been a technological explosion in the field of agriculture. This demands that the farmer has to know all aspects of technology prior to its adoption. It can only be possible through Information and Communication Technology.

The information and communication support during last 50 years has mainly been conventional; the extension personnel of the department of agriculture disseminated the technological messages to the farmers manually. This approach has not been able to reach majority of the farmers who are spread across the whole country.

This gap remains a challenge for the extension system even today. To reach over 110 million farmers, spread over 500 districts and over 6000 blocks is an up hill task. The diversity of agro-ecological situations adds to this challenge, farther farmer's needs are much more diversified and the knowledge required to address them is beyond the capacity of the grass root level extension functionaries.

Today it is possible to find a solution to this situation by using the potential of satellite based Internet technologies to meet the location specific information needs of the farmers. Information and communication networks are expanding very fast. The number of satellite based Internet connection in India has crossed the two million mark and the number of telephone connections is over 320 million. Internet connectivity has touched almost all the districts in the country and is moving down up to the village levels. Many pilot projects to connect rural community to cyber-space are underway at various locations. The initial response of the rural people, particularly farmers has been very encouraging; many organizations are trying to establish internet connectivity to make best use to satellite based communication technology.

2. STATEMENT / EXPLANATION OF THE PROBLEM

Nowadays various means for information communication are available. Those all means are very efficient, accurate and quick and somewhat cheaper in the field of disseminating the information from research stations to the farmers.

There is a large gap in terms of the information systems and technology in developing countries; however, there is a great need for the developing countries to get into the information society with lots of vigor because there is no way our country can develop without taking cognizance of participating in information highway, this needs promotion of internet services in support of integrated approach that can lead to the growth of vibrant rural and agricultural extension communication networks across nations, regions and the globe. An integrated approach provides rural people the benefits from communication networks that enable information to flow into and from rural communities and agricultural organizations. An integrated approach also fasters communication among the many intermediary organization that work for rural and agricultural development.

It is also true that Internet is not absolute solution for effective extension of rural and agricultural development but it does bring new information resources and can open up new communication channels for rural communication and agricultural organizations. It offers a means for bridging the gaps between developmental professionals, rural people and agricultural producers through the initiation of interaction and dialogue.

There has been a rapid increase in the use of the internet in developing countries, with regard to rural and agriculture development and the internal, extension organization that support the betterment of rural populations and improvements in agriculture have important roles to play for the benefits of improved communication and access to information. Thus to take a maximum benefits of the system this study was selected to know the expectations of the farmers towards Community Internet Centre at village level. By knowing this we could be able to establish the rural Internet centers on need based approach.

3. SCOPE OF RURAL INTERNET

Enormous benefits await rural communities and agricultural organization when communication improves between the non-government organizations, government services, private sector entities and educational institutes that support rural and agricultural development. By sharing information about their activities in the fields of agricultural, rural development, forestry, fisheries, health, nutrition and education these agencies can better serve rural people and farmers. They can make use of "Lessons learned", determines and use "Best practices" and coordinate information development approaches or successful development approaches. At the same time, rural communication and agricultural organizations can benefit equally from the vertical channels of communications that enable rural extension agencies and farmers to communicate with decision makers and other concerned with development.

An integrated approach to the expansion of Internet services will promote often neglected horizontal communication between agencies linked to rural and agriculture development. At the same time, an integrated approach will provide the tools enable rural people and farmers to enter into new vertical communication relationships with external agencies. Improving horizontal communication can improve the quality and relevance of information resources and physical resources available to rural people. Improving vertical communication between rural people farmers and decision-makers can improve the quality of decisions that affect rural communities and agricultural organizations.

An integrated approach provides for vertical communication by establishing rural Internet access sites, and by enhancing horizontal communication between such entities as agricultural development organizations, agricultural input and equipment suppliers, government extension services, rural development organization, health care agencies, agriculture research and documentation centres. Farmers can negotiate better prices from local buyers by using information on current market prices including national displayed on Internet. Internet services, in conjunction with existing and more widely used communication media such as rural radio, will enable the broadcast enhancement of information and communication resources for rural people.

4. SIGNIFICANCE OF THE STUDY

Experience of rural Internet users of many developing and developed countries indicate that the Internet provides them with a very convenient method for quickly accessing a large volume of information without being impeded by geographical barriers in the from of new ideas, discussion, expert advice, global understanding and cultural awareness that helps to make them better and more informed citizens. In addition to this, social benefits of indicating new opportunities to organize and advocate for social change, equalization of urban or rural disparities and new links between urban and rural communities were also experienced. Keeping all benefits in mind the present problem has been selected to know the expectation of the farmers towards ICT in agriculture at village level. The investigation found confronted with following questions on this issue.

1. What extent the farmers want to use Internet in their day to day life ?
2. Was there any influence of socio-economy and psychological characteristics of the farmers on their opinion about Internet?

With a view to find out answer to these questions the present investigation entitled, "Expectation of the farmers towards ICT in agriculture at village level" was carried out.

Objectives of The Study

1. To study the profile of the Internet facility expecting farmers.
2. To know the opinion of the farmers towards Internet facility and to study its relationship with their profile.
3. To study the expectations of the farmers regarding Community Internet Centre at village level.

5. REVIEW OF LITERATURE

Internet is a new communication system in India. Looking to the recent implementation of this facility in India, So far studies on exposure of Internet for farmers have not been carried out in India. Thus, for the present study support of some of the studies available on various sites of Internet have been studied and reviewed and are presented here.

5.1. Review on ITC for Agriculture

Moni (2003) reported that a journey "from digital divide to digital opportunities for sustainable development" has begun in the beginning of the 21st Century, and is becoming a powerful tool for fostering Agricultural Growth, Poverty Reduction and Sustainable Resource Use in India.

Mehta (2003) reported that Technological empowerment is very important for adding value to our mission of rural development. New technologies are vital for achieving continuous improvement in the way of farming and thus achieving economically and socially sustainable development. Extending the benefits of technologies to the smallest Indian farmer is perhaps a greater challenge then developing a technology. Our efforts in developing appropriate Information Technology should be based on what the Indian farmers needs and the question of how the technology shall be made adaptable by the farmer.

Rangarao (2003) reported that full and whole hearted support & co-operation from all research & developmental departments, both state & central, State of art investments in technology, infrastructure, people & network, Management's missionary zeal, commitment and dedication to the cause of Indian farmer and his welfare, dedicated network of professionals at ground level as well as the production centre at headquarters were the key factors for the tremendous popularity of EENADU and E-TV's ICT initiatives in agriculture.

Rathod (2003) reported that major applications of ICT for agriculture input and output marketing are Farmers Education, Fast and Effective Sales Promotion, Planning and Forecast, Logistic Support, Fast transfer of Data and Information, Net Working Tools for Global Linkage and Effective Decision making.

Sushma (2003) concluded that RS and GIS have significant role in Information for planning, monitoring Rural Development Programmes.

Shaik *et al.* (2003) reported that government have to reorganize their policy so that full fledged strategy is developed at all level they also pointed out that research should be carried out to develop new research paradigm for ICT in agriculture development.

Chaudhary, *et al.* (2003) disclosed that farmers make on farm decisions based on technological information about crop production along with information of farms owned him/her, past history of crops cultivated, agro-climatic conditions, availability of agro-inputs, economical condition, market prices, etc. The lack of availability of proper information restricts farmer to utilize his capability to become more innovative and productive in various aspects of farming thus, suggested Crop Production Information System (CPIS) to aid farmers to make correct and timely decisions regarding farm activities. The system design of CPIS is structured to incorporate context based knowledge and information regarding farm production system. CPIS could support retrieval of required information pertaining to a relevant farm situation.

Patel and Patel (2003) concluded that the Internet is a tool having potential to contribute to agricultural development as one can access to a vast global information resources. It enables communities for two-way communication. It can facilitate dialogue among communities, planners, development agencies, researchers and technical experts & encourages community participation in decision making.

Sontakki (2003) concluded that NAARM has recognized IT as the core tool to empower the agricultural professionals to successfully meet the complex challenges of the future. Towards this end, NAARM has identified a road map to meet the growing needs of IT and Computer Applications in various aspects of agricultural research, education and extension management. This covers initiatives like develop a comprehensive IT policy for the Indian NARS, Status paper on Indian Experiences on ICTs for Agriculture and Rural Development, Training and HRD efforts to be intensified to cover the emerging areas and focusing on ICAR/SAU/Private R&D and Agri-business Agencies, IT and Computer courses to promote Office Automation in NARS Institutions, Micro-level Research on taking ICTs to farmers, e-learning and Web-based Education and Training, Advanced Decision Support Systems, Strengthening VLC and Impact assessment of ICT initiatives in Indian agriculture.

Surya and Amardeep (2003) concluded that New ICTs have a major role to play in the life of farmers as they provide them latest know-how on agriculture, on–line selling and buying. They can also access daily weather forecasts, information on cropping patterns, soil conservation, and government schemes. These developments have also opened the door to a whole new generation of 'multi-modal' market information services, which bring much needed price and product information directly to farmers. India's advantage lies in its population density, its network, and economies of scale, understanding of rural poor needs and its cultural and geographic diversity. By addressing these challenges and connecting up rural communities, India can boost its efficiency and economic competitiveness. The world economy is moving very fast and the last thing India can afford is to leave most of its population disconnected from the high-speed global knowledge network. India would benefit from approaching the issue by using ICT in Agriculture by focusing on providing broadband connectivity and a centric development approach.

Aditya (2003) reported that coming of knowledge society and making of information a commodity. Knowledge Management is the key for business, household, agriculture, conservation and sustainability.

The connectivity, low cost computer for masses, sustainability, computer Education and E-Governance are the key issues to enable an ICT based model to rural India.

6. OTHER RELEVANT REFERENCES

Andrew Trotter (2002) reported that Internet access has a measurable impact on student achievement. The researchers specifically looked at the impact of the federal Education (E)-rate program, a program designed to help schools acquire telecommunications services. Though the results do not indicate that Internet investment has made an impact, critics of the study suggest that not only is it too early to make this kind of assessment, but that Internet connections alone are not likely to improve student test-scores. Nevertheless, the study serves as a "First crack" at trying to determine if this link between Internet connections and student achievement can be made and it provides some interesting conclusions.

Anonymous (2001) nearly 25 per cent of professionals who use IM (Chatting) at home have started using it to contact clients and colleagues, according to a Media Matrix survey conducted in April.

Anonymous (2002) Assessment Bulletin Volume 6, Number 1 November 2002. In this issue reported on the results of a survey of UMass Amherst freshmen and sophomores about their exposure to course-based instructional technologies, including E-mail, course websites, the Internet, video, computer-generated displays, and computerized quiz and response systems. Students were also asked about their interest in online courses. The survey found E-mail is commonly used for communication between professors and students, course websites and required Internet research are also fairly common, students tend to be exposed to course-based interactive technologies less frequently and student interest in taking online courses is mixed; few have actually taken such a course.

Anonymous (2002)* reported that Information technology has been one of the most aspired fields in today's world. Integrating IT with agriculture will help any country to regulate its overall economy and trade. The different IT Technologies like Expert System in Decision support system, Remote sensing, Gyandoot etc., have brought revolution in Indian agriculture. Even corporate like ITC, MSSL, HLL, are now looking forward to extract huge benefits out of this collaboration of IT with agriculture.

Anonymous (2004) reported that Internet-based technologies are changing logistics from being a packaging and moving function into an information business. Electronic business integrates carriers with shippers via electronic ordering, inventory decisions and product flow. Retailers and third-party logistics firms are able to access their respective information systems through the Internet.

Barbara A. Walczak (1996) concluded that the more students are exposed to what the Internet has to offer, the better and more interested users they become.

Campbell *et al.* (2001) reported that eventhough individuals express concern and awareness about Internet privacy, they were still willing to engage in risky online activities. Males expressed more concern and awareness about Internet privacy, yet they engage in more risky behaviors than did females.

Curtis (2002) reported that the age of Internet user respondents was quite varied. Nearly one-fourth of the participants were under age 36, half were 36 to 50 years old, and slightly more than one-quarter were over age 50. He further reported that the pool of respondents was fairly well educated. In terms of highest degree held, 3 percent had high school diplomas, 8 percent obtained some type of professional certification beyond high school, 35 percent possessed bachelor's degrees, 41 percent had master's degrees, 8 percent held advanced degrees or were ABD, and the remaining 5 percent had earned a doctoral degree. In addition to this he also trued to find out purposes of online use as expressed by the respondents. Most of were using Web-based learning as an alternative to instructor-led courses (66 percent) or as a supplement to traditional instructor-led training courses (53 percent). About one-fourth used it as a follow-up to live instruction. One in five used the Web as the sole source for learning. The most prevalent online skills learned were computer applications and software skills (64 percent) as well as technical skills (50 percent), the next most popular type of online training concerned job-related skill development (45 percent). Communication skills, computer systems or programming skills, and management or supervisory experience were all offered at roughly 3 in 10 organizations surveyed. Around one-fourth offered online training for personal growth as well as customer service skills. They tended not to use the Web for sales or marketing skills (16 percent) or executive education (13 percent).

Goh (1997) found that friends and family members were the key influencing factors that encouraged use of Internet. However, the high cost of computers and computer peripherals was the main deterrent for non-users. Lee (1997) also discovered that about 26 percent of his respondents did not have basic computer skills, but showed high interest in learning and using the Internet. Pang (1997) noted, however, that pornography and not business, knowledge or communication, motivated the use of Internet among his respondents.

Goh (1997) identified several factors that influenced the use of Internet in the Klang Valley, namely, education, affordability, and the need for infotainment. Education and a positive attitude toward IT were necessary prerequisites for IT acculturation. In the three studies, it was found that largely academicians and professionals used Internet.

Mayes (1995) studied differences in attitude change, towards the Western Civilization class, for students using web-based resources and those using traditional textbooks. In this study no differences were found in change of attitude towards the Western Civilization class between the students that used web-based resources and students that used traditional textbooks. These findings agree with those of Gee (1990) who also found no differences. However, the study disagrees with the findings of (Koohang, 1989 and Rothman, 2000) who found differences in attitude change. As acknowledged by Erwin (2001), these contradictory findings could be attributed to the complexity of measuring attitude. Erwin (2001) insists that attitude is an important construct and must be measured because of its usefulness in predicting behavior. Further explanation for a lack of attitude change could be supported by the findings of Gee (1990) who reported student's perception was affected by the fact they found Internet to be time consuming.

Richard C. Sherman (1997) reported that, The World Wide Web (WWW) is an exciting new tool for teaching college courses in psychology. The potential benefits of the WWW stem from the wealth of information it makes available to instructors and students, the ease of access to that information, and the "hypermedia" richness of WWW documents.

Sharon Gudin (2003) reported that American companies lead significantly in the strategic use of IT, according to a study by Info-Tech Research Group, a technology research firm. The study shows that 30 per cent of Americans reported that technology is used as a `strategic weapon' within their organizations compared to 17 per cent in the U.K. and 6 per cent in Canada.

Teresa *et al.* (2003) indicated that when question asked students during their survey to specify their level of comfort when seeking information from a variety of sources, including search engines, a library Web page, a friend, a professor/TA/GSA, a faculty or class Website, or the library. Overwhelmingly, students responded that they were most comfortable seeking information from a search engine, with 91.45 per cent selecting "very comfortable" or "comfortable." The search engine receiving the most frequent mention was www.google.com.

7. RESEARCH METHODOLOGY

7.1 Location of Research Area

Present study was undertaken in Anand District of Gujarat because of the following reasons.

1. This District of Gujarat the potential in agriculture and animal husbandry.
2. The farmers of this district are very enthusiastic in adoption of new technology, latest facilities and modern equipments in their daily life.
3. As compared to other Districts of the state, farmers have more connections with any modern media as opportunity of first rural farm TV station of India at village Pij was given to the farmers of this district.
4. As compared to other districts of the state, there is a tendency to go abroad among the members of the farming community of this districts, thus farmers have link with their relatives living in foreign countries through modern media like Internet.
5. As compared to other Districts, farmers are more active in getting best results from the agriculture and animal husbandry.

6. The head quarter of Anand Agricultural University (AAU) is situated in this district of the Gujarat is doing research and extension activity to popularize agricultural and animal husbandry related technique in various crops in addition to this AAU is also trying and supporting State Government to strengthen internet facility for the farming community.
7. This District is one of the potential districts of the Gujarat state; where due the availability of canal irrigation, socio-economic condition is better than other districts and farmers are having mentality to use modern scientific facility to uplift their life.
8. So far there is no any study has been carried out in this district to understand expectations of the farmers about the use of Internet facility for extension work.

7.2 Sample Size

Present study was undertaken in purposively selected Anand taluka of the Anand district of Gujarat. As study was to know general expectations of the farmers regarding use of modern facility of internet, four random villages e.g. Vasad, Mogar, Navli and Napad having agriculture as major source of income and are known for their progressiveness with more than 5000 populations and comparatively good infrastructure facilities in villages were selected purposively. Total 25 respondents having farming as a major source of income from the each village were selected randomly. Thus, a random sample of total 100 farmers from purposively selected four villages was selected for the study.

7.3 Research Design

As the study was concerned to understand the expectations and opinion of the farmers regarding use of Internet for extension work, ex-post-facto research design was used for this study. This design was used because the researcher did not have any control on the independent variables of the selected farmers.

Ex-post-facto research design is systematic experimental investigation in which control on independent variable is very weak and in this study all most no control was possible. Kerlinger (1976) stated that ex-post-facto research design is worthy to apply when the independent variables have already acted up on.

7.4 Derivation of Hypothesis

On the basis of objective of the study, the following null hypothesis was formulated.

Ho-there will not be any relationship between selected characteristics of farmers and their level of opinion regarding use of internet for extension work.

7.5 Limitation of the Study

The following are the limitations of the study.

1. The area of present study is restricted to only 100 farmers of the district of all the farmers.
2. Though all possible precautions were taken to make the study precise, objective and reliable, the study was restrained only to Anand and Ankalav Talukas of Gujarat because of limited time and resources.
3. The study is limited to only selected characteristics of the respondents and not all the characteristics.
4. The findings will be based on the verbal expression and response of the respondents.

7.6 Method of Data Collection

Data were collected by personal contact method. The tool used for study was personal interview schedule. Keeping in view the objectives of the study, the interview schedule was developed. In formulating questions and statements for schedule the investigator referred the reviewed, related literature, popular articles, and research reports and consulted the major advisor to seek their opinions and suggestions to make interview schedule more scientific and meaningful and accordingly it was developed. The respondents were interviewed at their home or at the community place. Before conducting interview the respondents were explained about the purpose of the study. The questions from the schedule were asked one by one and their responses were recorded on the spot. Every possible care was taken to maintain congenial atmosphere for getting unbiased responses from the respondents.

7.7 Selection of the Variable

The variables under study were selected on the basis of extensive review of literature related to the subject and consultation with experts and finally the variables that are found to be most relevant to the present study are selected as under.

7.7.1 Dependent Variable

1. Expectation of the farmers regarding use of Internet for agricultural extension.

7.7.2. Independent Variables

A. Personal
   1. Age
   2. Education

B. Economic
   1. Occupation
   2. Land holding
   3. Animal wealth
   4. Possession of modern equipment

C. Social and Communicational
   1. Type of family
   2. Caste
   3. Relatives in aboard
   4. Mass media exposure
   5. Social participation

D. Psychological
   1. Opinion of the farmers towards Internet

7.8 Measurement of Variables

7.8.1 Measurement of expectation of the Farmers towards Community Internet Center (CIC) at Village Level

1. Expectation of farmers to have Community Internet Centre at village level
2. Suggested place for CIC
3. Expectation of services at CIC
4. Choice of person to manage CIC
5. Choice to take initial financial support to start CIC
6. Choice to bear expenses to run CIC

7. Choice to have minimum Computer set at CIC
8. Expectation about the form of information on internet and at ICT
9. Purposes to have CIC
10. Expectation to involve in the process of exploring internet at CIC
11. Expected frequency of the farmers to use internet at CIC
12. Expected agency to receive training to use CIC.

7.8.2 Measurement of Independent Variables

7.8.2.1 Age

Age of the respondents was measured in completion of year of respondents at the time of interview. The respondents were classified into three groups, viz., young (18 to 31), middle age (31 to 55 years) and old age (above 55 years).

7.8.2.2 Education

Level of education of the respondents was measured in number of formal education in year completed by the respondents and they were classified in to three group viz., primary, high school and higher secondary, above higher secondary.

7.8.2.3 Occupation

It refers to involvement, connection and attachment of the respondents in various income generating activities. The respondents were classified in three groups, farming only, farming + A.H, and farming + service.

7.8.2.4 Landholding

Landholding refers to the actual area of land possessed by the respondents based on landholding the respondents were categorized into three groups viz. Small size of land holding (up to 2 ha.), Medium size of landholding (2.1 to 4 ha.) and Large size of landholding (Above 4 ha.). One score was given to each hectar of land to quantify this variable.

7.8.2.5 Animal Wealth

It refers to total number of animals possessed by the farmers. The respondents were classified in four groups, viz., No animal, Up to 2 animals, 3 to 5 animals and Above 5 animals. One score was given to each of the animal possession to quantify it.

7.8.2.6 Possession of Modern Equipment

This variable was measured in terms of implement or equipment possessed by the respondents. The scores of 10, 1, 2 and 3 were given to the possession of tractor, thresher, power tailor and tractor mounted sprayer, respectively.

7.8.2.7 Type of Family

The type of family is most important among social factor. It was measured with the help of SES scale developed by Pareek and Trivedi (1963). The respondents were classified into two groups like Nuclear family and Joint family and score of one and two were given to quantify this variable.

7.8.2.8 Caste

Caste of the respondents was divided into three categories and measured with score of three was assigned to farmers from higher caste, two to farmers from intermediate caste and one to the farmers from lower caste.

7.8.2.9 Relatives in Aboard

It refers to possession of any relatives by the respondents in abroad countries. The respondents were classified in to two groups, viz., no relatives and relatives in abroad countries. One score was given to the respondents with relatives and zero to the respondents without relatives.

7.8.2.10 Mass Media Use

It is defined as the nature and Number of respondents' involvement with mass media like television, radio, news paper, internet and others. It was measured in terms getting responses of the respondents in four categories viz., everyday contact, once in week contact, fortnightly contact, more than fortnightly contact and never contact and scores of 4, 3, 2, 1 and 0 were assigned to them, respectively.

7.8.2.11 Social Participation

Social participation refers to the degree of involvement of respondents in formal organization either as a member or office bearer. The social participation of respondents was qualified on the basis of SES scale developed by Trivedi (1963). The respondents were classified into three groups like no membership (0 score), Membership in one organization (1 score) and Membership in more than one organization (2 score).

7.8.2.12 Opinion of Farmers towards the use of Internet

The level of opinion of the farmers about Internet was measured with the help of preparing ten statements by consulting major guide and experts of extension personals and information technology centre of Anand Agricultural University, Anand. The responses of the respondents were obtained against each statement in terms of their agreement or disagreement on a three point continuum viz., 'agree', 'partially agree' and 'disagree' with scores of 2,1 and 0 for positive statements, whereas scoring was reverse for negative statements. Total score was obtained by summing all the responses. High score indicated highly favorable opinion, while low score indicated negative opinion of the farmers towards Internet.

7.9 Statistical Framework for the Analysis of Data

The following statistical tools were used for interpreting the data.

7.9.1 Number and Percentage

Simple comparisons were made on the basis of Number and percentage.

7.9.2 Mean

The mean was obtained by the total number of score divided by total number of trainees.

7.9.3 Coefficient of Correlation

Coefficient of correlation was computed to find out the relationship between each of the independent variable and the dependent variable by employing following formula.

$$r = \frac{n\Sigma xy - \Sigma x \, \Sigma y}{\sqrt{[\Sigma n\Sigma x^2 - (\Sigma x)^2]\,[\Sigma n\,(\Sigma y) - (\Sigma y)^2]}}$$

Where, r^2 = correlation coefficient
x = Independent variable
y = Dependent variable
n = Total number of respondents

8. RESULTS AND DISCUSSION

This portion of the assignment presents the objective wise findings of the study. Keeping in view the objectives of the study, information was collected from the respondents and analyzed. The results regarding the following aspects are presented in this chapter.

8.1 The profile of the Internet Facility Expecting Farmer

8.1.1 Age

The data presented in Table 1, indicate that majority (64.00 per cent) of the CIC expecting farmers had middle age, followed by 29 per cent with young and only 7.00 per cent with old age. Thus it can be concluded that majority of the young and middle aged farmers were from those the CIC expecting farmers. The reason for this might be that farmers young and middle age group have awareness regarding significance of Internet facility at village level. The another reason might be that the parental occupation taken up by middle age groups might have realized role of CIC for getting high tech oriented agricultural information to develop their agriculture and lifestyle.

Table 1 : The respondent according to their age

N=100

Sr. No.	*Age group*	*Number*	*Per cent*
1	Young (18 to 31 years)	29	29.00
2	Middle age (32 to 55 years)	64	64.00
3	Old age (above 55 years)	07	7.00
	Total	100	100.00

8.1.2 Education

It is evident from the data in Table-2 that slightly less than half (45.00 per cent) of respondents had high school and higher secondary school level of education, followed by 38.00 per cent with up to primary level of education and 17.00 per cent with above higher secondary level of education. From this fact, it can be concluded that the interest to have CIC at village level was observed among low and highly educated farmers.

Table 2 : The respondent according to their level of education

N=100

Sr. No.	*Level of education*	*Number*	*Per cent*
1	Primary level	38	38.00
2	High school and higher secondary	45	45.00
3	Above higher secondary school	17	17.00
	Total	100	100.00

8.1.3 Occupation

The data in Table 3 indicate that majority of respondents had farming and animal husbandry as their main occupation, followed by 28 per cent with only farming as their sources of livelihood and 14.00 per cent with farming and service as their source of livelihood. The results indicate that majority of respondents were found to be dependent on farming with animal husbandry.

Table 3 : The respondents according to their occupation

N=100

Sr. No.	*Occupation*	*Number*	*Per cent*
1	Farming	28	28.00
2	Farming + A.H	58	58.00
3	Farming + service	14	14.00
	Total	100	100.00

8.1.4 Size of Landholding

The data presented in Table 4 indicated that majority (60.00 per cent) of the respondents had 2 hectare of land holding, followed by 26.00 per cent with 2.01 to 4.0 hectare and 14.00 per cent with above 4 hectare of landholding. It can be concluded from above finding that majority of the respondents had landholding below 2 hectare.

Table 4 : The respondent according to landholding

N=100

Sr. No.	*Landholding*	*Number*	*Per cent*
1	Small (up to 2 ha.)	60	60.00
2	Medium (2.01 to 4 ha.)	26	26.00
3	High (above 4 ha.)	14	14.00
Total		100	100.00

8.1.5 Animal Wealth

The data in Table 5 indicate that nearly half of respondents (46.00 per cent) had up to 2 animals followed by 24.00 per cent with 3 to 5 animals, 02.00 per cent with more than five animals and 28.00 per cent had no animal.

Thus it can be said that majority of respondents had up to 2 or more than two animals. Dairy is well developed and marketing is not a problem in this taluka, thus to earn additional income majority of the farmers prefer to have animal husbandry with agriculture.

Table 5 : The respondents according to their animal wealth

N=100

Sr. No.	*Animal wealth*	*Number*	*Per cent*
1	No animal	28	28.00
2	Up to 2 animals	46	46.00
3	3 to 5 animals	24	24.00
4	Above 5 animals	02	02.00
Total		100	100.00

8.1.6 Possession of Modern Equipments

The data in Table 6 indicate that 46.00 per cent, 22.00, 18.00 and 14.00 per cent of the farmers had equipments likes tractor, power tailor, tractor mounted sprayer and thresher, respectively.

Table 6 : The respondents according to tenure of modern equipments

N=100

Sr. No	*Modern equipment*	*Number*	*Per cent*
1	Tractor	46	46.00
2	Thresher	14	14.00
3	Power tailor	22	22.00
4	Tractor mounted sprayer	18	18.00
Total		100	100.00

Thus it can be said that slightly less than one fourth (46.00 per cent) of the respondents were having tractors. As seen in this study, majority of the respondents had small size of landholding thus, they might not have required costly equipments with in established cropping pattern in such a small size of landholding.

8.1.7 Type of Family

The information furnished in Table 7 reflects that majority (62.00 per cent) of respondents had joint type of family, while 38.00 per cent of the farmers had nuclear type family.

Table 7 : The respondent according to their type of family

N =100

Sr. No.	*Type of family*	*Number*	*Per cent*
1	Nuclear family	38	38.00
2	Joint family	62	62.00
Total		100	100.00

Result indicates that yet in the family of the farmers, joint family system is existed.

8.1.8 Abroad Relationship

The data in Table 8 indicate that majority (90.00 per cent) respondents have no relation in abroad and follows by 10.00 per cent respondents had relation in abroad.

Table 8 : The respondents according to their relation in abroad

N=100

Sr. No	*Relation in abroad*	*Number*	*Per cent*
1	No relation	90	90.00
2	Relation with	10	10.00
Total		100	100.00

It was striking to note that even in the family of the farmers also at least ten percent of them had either their son, brother, daughter or any close relatives are staying in abroad countries. It shows mental progress or forwardness in Gujarati farmers and this is one of the reasons why farmers of the state expect modern facility of communication in rural areas also. It was also seen during the course of investigation that majority out of remaining 90 per cent farmers were also expecting to send their son or daughter in abroad countries. But over and above it can be concluded from above findings that majority of the respondents were having no relation in abroad.

8.1.9 Mass Media Use

The perusal of the data in Table 9 revealed that the television was used by highest respondents followed by radio, news paper and other, respectively. Here, it is seen that nowadays farmers were using Internet in a very low level but it does not mean that they don't want to use it. In fact due to unavailability of community Internet centre they were not in position to use this facility. It should be noted here that in this study it has been observed that cent percent of the farmers had expectations to have Internet facility in their villages.

The data in the same table conclude that the television was the most used mass media in rural area.

Table 9 : The respondents according to use of mass media

N=100

Sr. No.	*Mass media*	*Mean value*	*Rank*
1	Television	2.90	I
2	Radio	1.92	II
3	News paper	1.78	III
4	Others	0.20	IV
5	Internet	0.10	V

8.1.10 Social Participation

Data presented in Table 10 indicated that the majority of the farmers (55.00 per cent) had membership in one organization followed by no membership of any organization (40.00 per cent) and only 5.00 per cent of the farmer had membership in more than one organization.

Thus it can be concluded that majority (60.00 per cent) of the farmers had membership in one or more than one social organization. The possible reason for this result might be that the milk co-operative is actively working at village level in this area of the state in most of the villages. So that membership in at least one organization was commonly observed.

Table 10: The respondents according to social participation

N=100

Sr. No.	*Type of social participation*	*Number*	*Per cent*
1	No membership	40	40.00
2	Membership in one organization	55	55.00
3	Membership in more than one organization	05	05.00
Total		100.00	100.00

8.2 The Opinion of the Farmers Towards Internet

For all round development of Indian agriculture and farming community, need of Information Technology is realized by administrators, planners and Government in India. The Governments of many states in India have started their efforts in this direction. Before making this marvelous system popular among farming community, it is necessary to collect the opinion of the farmers towards this innovative system. An opinion is feeling, reaction or sensitivity of any individual towards particular object, technology or idea. The positive opinion plays an important role in making any technology acceptable among its real users. Looking to this fact opinion of the farmers towards Internet was studied and the results are presented in Table 11.

Table 11: The opinion of the farmers towards Internet

N=100

Sr. No.	*Statements*	*Opinion*		
		Agree per cent	*Partially agree per cent*	*Disagree per cent*
1	Internet can be a very useful mean to the farmers during present time	69.00	10.00	21.00
2	Internet is rich source to collect worldwide information on agriculture and allied fields	71.00	12.00	17.00
3	Internet is fastest way to exchange information in shortest time	72.00	14.00	14.00
4	Internet is costly affair for the farmers	41.00	24.00	35.00
5	Internet is best mean to collect information on market prizes of agricultural product	41.00	35.00	24.00
6	The use of Internet is nothing other than time pass activity	28.00	21.00	61.00
7	Information available on Internet is easy to understand	12.00	41.00	47.00
8	Development of Indian farmers is possible through internet	56.00	30.00	14.00
9	I wish that my children should make positive use of internet facility	81.00	10.00	09.00
10	I wish that farmers should make use of internet	81.00	19.00	00.00

In order to know the opinion of the farmers towards Internet, total ten statements were prepared and responses were collected in terms of there continuums viz., 'Agree', 'Partially agree' and 'Disagree". It is motivating to know that 71.00 per cent of the farmers understood that internet is a rich source to collect world wide information on agriculture and its allied fields, while 72.00 per cent supported to the statement that 'Internet is fastest way to exchange information in shortest time'. Looking to the present cost involved in this technology, it is not easy for the farmers to have this facility individually at his home, so 65.00 per cent of the farmers completely or to a certain degree felt that Internet is costly affair for the farmers. Internet is best mean to collect information on market prizes of agricultural products but as it is being new system for our farmers, mix opinion was observed for this aspect and it was observed that 41.00 per cent of the farmers realized its use for agricultural marketing while 35.00 per cent partially realized it and nearly one fourth (24.00 per cent) of them did not realize this feature of internet.

The results point out that majority (69.00 per cent) of the respondents agreed with the statement that "Internet can be a very useful mean to the farmers during present time". During the study it was exciting to note that majority (61.00) per cent of the farmers did not believe that use of internet is only time pass activity. It means that they opined it as a useful medium for farming community.

Internet is such a smart technology that a variety of information can be collected with in few seconds by any person from the any corner of the world. Internet is also a great source for agricultural information. In spite of this Indian farmers are not in position to use this information for the development of their agriculture because whatever sites available on agriculture are in English language. This fact was reflected in study so 88.00 percent of the farmers partially or absolutely realized that information available on Internet is difficult to understand.

It was really pleasing to note that majority (86.00 percent) of the farmers out of them 56.00 per cent totally and 30.00 per cent partly assumed that development of Indian farmers is possible through Internet. Internet is best mean to learn all now things for young generation, so 82.00 per cent of the farmers were wishing their children to make positive use internet at the same time 81.00 per cent of them had opinion that farmers should make use of internet.

8.2.1 Relation between Profile of the Farmers and their Opinion about Internet

8.2.1.1 Age and Opinion

Data presented in Table 12 indicated that age of the respondent was observed positive with their opinion about Internet. It means that old aged farmers had favorable opinion regarding Internet as compared to young age farmers. In fact this trend was not up to the level of significance. In fact trend of relationship says that old aged farmers during their traditional way of containing extension personnel might have traced difficulties. In such a situation they might have expected better results through the use of Internet for agriculture extension. But because the statistical "r" value indicated that there was non-significant relation between age and opinion of the respondents regarding the use of Internet, thus the null hypothesis (HO), in case of age was accepted.

Table 12: Relation between profile of the farmers and their opinion about internet

N=100

Sr. No	*Independent variables*	*Correlation coefficient ("r" value)*
I	**Personal variable**	
1	Age	0.1368
2	Education	0.3263*
II	**Economic variable**	
3	Occupation	-0.0092
4	Landholding	0.2459*
5	Animal wealth	0.0195
6	Modern agricultural equipment	-0.0087

contd...

Sr. No	*Independent variables*	*Correlation coefficient ("r" value)*
III	**Social and communication**	
7	Type of family	-0.2472*
8	Contact with NRI's	0.2603*
9	Experience of internet use	0.2718*
10	Mass media exposure	0.2462*
11	Extension contact	0.1239
12	Organization participation	0.1247

* Significant at 0.05 level of probability

8.2.1.2 Education and Opinion

Level of education of the farmers was positively significant with their opinion regarding Internet. Results reveal that opinion towards Internet was found more positive with incase in the level of education of the farmers. High level of education might have given opportunity to farmers to understand value of Internet to collect useful information where and when required. Hence, the null hypothesis (HO) in relation to education was rejected and it can be said that there was significant relationship between the education of the farmers and their opinion about Internet.

8.2.1.3 Occupation and Opinion

The result shown in Table 12 reflects that there was negative relation between opinion of the farmers regarding use of internet for communication and their involvement in different occupations. It indicates that opinion about internet use was observed more positive among those farmers who were involving in less number of occupations or only in agriculture occupation. It is natural that when farmer is dependent only on agriculture, he will try to be more positive in all those matter through which he can get more economic return. Internet is such a good technology through which farmers can get better agricultural production. In fact this trend was up to the noticeable level thus; the null hypothesis (HO) in case of occupation was accepted.

8.2.1.4 Land Holding and Opinion

The results revealed that size of landholding by the farmers was observed positive significant with their opinion regarding Internet use. It means that more positiveness in opinion towards Internet was observed among the farmers who had bigger size of landholding. The farmers with big size of landholding will have comparatively better economic condition and have mentality to own all those materials and facilities which make their life more meaningful, happy and prosperous. Thus farmers with big size of landholding will have positive feeling for such kinds of materials and facilities. This might be the reason to have positive relationship between the landholding and opinion of the farmers about Internet. Therefore, the null hypothesis (HO) in relation to landholding was rejected.

8.2.1.5 Animal Possession and Opinion

The statistical "r" value presented in table shows that there was non-significant relationship between opinion of the farmers regarding use of Internet and number of animal possessed by them. It means that degree of positiveness in term of opinion toward Internet use was observed homogenous among the farmers who had small, medium and big size of animal possession. In other words it can also be said that all kinds of animal possessors had similar kind of positivism towards Internet. Hence, the null hypothesis (HO) was accepted in this case too.

8.2.1.6 Possession of Modern Agricultural Equipments and Opinion

Here also non-significant relationship was observed between possession of modern agricultural equipments and opinion regarding use of Internet. In other word it can be said that possession of more or less modern agricultural equipments by the farmers had not played any significant role in improving there opinion regarding use of internet. It is therefore null hypothesis (HO) in this case was accepted.

8.2.1.7 Type of Family and Opinion

Data presented in the table 12 shows that there was negative significant relationship with type of family of the farmers and their opinion towards Internet use. It means that more positive opinion was observed among those farmers who were having nuclear type of family. It is obvious that in nuclear type of family, head of the family is from younger age group. The young aged persons are more enthusiastic and ready to accept new changes and modern facilities in the life. This might be the reason to have more positive opinion of the farmers who were from nuclear family. Hence, null hypothesis (HO) in case of type of family was rejected and above conclusion was drawn.

8.2.1.8 Contact with NRI Relations and Opinion

It was assumed that farmers having their relatives in foreign countries would have more knowledge regarding Internet use for communication purposes. Thus, relationship between contact of farmers with NRI relatives and their opinion regarding Internet use was worked out. Positive "r" value indicates that there was significantly positive opinion towards Internet use among those farmers who had their relatives settled in foreign countries. The farmers having their relatives in foreign countries might have understood importance of Internet while communicating with them. Thus null hypothesis (HO) in this case was also rejected and it was concluded that there was significant relationship between the contact of the farmers with NRI relatives and their opinion about Internet.

8.2.1.9 Experience of Internet use and Opinion

A result presented in Table 12 indicates that there was positive relationship between the experience of Internet use and opinion of the farmers about Internet use. It means that opinion of those farmers was observed more positive, who had experience of communicating via Internet to their NRI relatives. It means that farmers with pleasant experience and familiarity of Internet had understood its real values, thus showed their positive opinion towards the use of Internet. Here also null hypothesis (HO) was rejected and above conclusion was drawn.

8.2.1.10 Mass Media Exposure and Opinion

The "r" value presented in table 12 indicates that level of opinion towards Internet use was more positive among those formers who had more exposure with mass media, viz., television, radio and newspaper. Farmers during their exposure with all such media might have understood importance of Internet for speedy communication. This might be the reason for positive significant relationship between mass media exposure of the farmers and their opinion regarding Internet use. Therefore, null hypothesis (HO) was rejected and above conclusion was drawn.

8.2.1.11 Extension Contact and Opinion

The data presented in the Table 12 shows that, there was positive relationship between extension contact of the farmers and their opinion regarding Internet use in agriculture extension. In fact this tendency was not up the level of significant. It means that extension personnel had played a little role in improving opinion of the farmers regarding Internet use but it was not up to the level of significant. It is therefore the null hypothesis (HO) was accepted and it can be concluded that extension personals have not shown any effect in improving the opinion of the farmers towards Internet.

8.2.1.12 Organization Participation and Opinion

The result presented in the Table 12 shows that relationship between organizational participation of farmers and their opinion regarding internet use was positive but non significant. It means that opinion regarding internet use was observed more positive among those farmers who had more organizational participation, but this propensity was not up to the level of importance. It is therefore null hypothesis (HO) was accepted and it was conformed that there was non significant relationship between organizational participation of farmers and their opinion regarding internet use.

Thus from the result of the Table 12, it can be concluded that Education, Landholding, Contact with NRI's, Experience of internet use and Mass media exposure were observed positively significant while, type of the family was observed negatively significant with the opinion of the farmers about the use of Internet for farming community. The independent variables like Age, Occupation, Animal wealth, Modern agricultural equipment, Type of family, Extension contact and Organization participation were observed non significant with the opinion of the farmers about the use of Internet for farming community.

8.3 The expectation of the Farmer Regarding Community Internet Centre (CIC) at Village Level

For the various aspects to have Community Internet Centre (CIC) at village level, information was collected from the farmers to know their expectations. The results are given in tables below.

8.3.1 Expectation of Farmers to have Community Internet Centre at Village Level

Expectation of the farmers to have Community Internet Centre (CIC) was measured in terms of two responses viz., 'Yes' and 'No'. Table 13 indicates that cent percent of the respondents expected to have CIC at village level. It shows that farmers have realized importance of Internet facility in villages as an effective source of information as well as for the speedy communication.

Table 13: The expectation of CIC at village level

N=100

Sr. No	*Type of Expectation*	*Number*	*Per cent*
1	Yes	100	100.00
2	No	00	00.00
	Total	100	100.00

Thus, it can be concluded that cent percent (100.00 per cent) of the respondents expected CIC at village level.

8.3.2 Suggested Place for CIC

When Internet facility is made available at community base, choice of the place of community Internet center should be known from its real users. Here the choices of the places were known for many public places. The respondents have suggested more than one place for the same. It can be observed from the data in table that Building of Panchayat was preferred by 95.00 per cent of the respondents, followed by building of primary school by 81.00 per cent, building of co-operative dairy by 76.00 per cent and buildings of high school and community hall by only 5.00 and 3.00 percent respondents, respectively. The building of panchayat is such an informal place, where people feel much familiarity, thus it was preferred by great majority of the farmers to have CTC.

Table 14: Respondent according to choice of the place of CIC

N=100

Sr. No	*Place*	*Number*	*Per cent*
1	Panchayat	95	95.00
2	Primary school	81	81.00
3	Co-operative Dairy	76	76.00
4	High school	05	05.00
5	Community hall	03	03.00

Thus it can be concluded that slightly less than cent per cent (95.00 per cent) of the respondents preferred panchayat office as the best place of CTC.

8.3.3 Expectation of Service at CIC

The services and facilities provided at CTC plays a major role in it success thus; expectations of respondents in this direction were also studied. The data in Table 15 indicate that operator or guide, printer, separate cabin, downloaded information in printed form and extra seating facilities were expected by 98.00, 72.00, 68.00, 63.00 and 53.00 per cent of the farmers, respectively.

Table 15 : The respondents according to expectation of service

N=100

Sr. No	*Expectation of service*	*Number*	*Per cent*
1	Operator/Person to Guide and help	98	98.00
2	Printer	72	72.00
3	Separate cabin	68	68.00
4	Collected information in printed form	63	63.00
5	Extra seating facilities	53	53.00

Thus, from the above findings it can be said that majority of the farmers were expecting operator or guide, printer, separate cabin, downloaded information in printed form and extra seating facilities at CIC.

8.3.4 Choice of Person to Manage CIC

Table 16 : Respondents according to choice of person to manage CIC

N=100

Sr.No	*Person*	*Number*	*Per cent*
1	Agriculture graduate with computer knowledge	98	98.00
2	Educated person of village	76	76.00
3	Expert of computer	57	57.00
4	Primary school teacher	28	28.00

The proper guide and manager of CIC will always run the centre effectively. With a view to collecting proper choice of the farmers for the person as a manager of CIC, information was collected. While giving their choices many farmers gave more than one choices also. The data in Table 16 indicate that agriculture graduate with computer knowledge was preferred by nearly cent per cent (98.00 per cent) of the farmers as a manager of the centre, followed by any educated person of the village was preferred by 76.00 per cent, expert of computer by 57.00 per cent and primary school teacher by 28.00 per cent of the farmers. It means that farmers have feelings to get advantages of internet not only for communication but also for having information on agricultural subjects, thus agriculture graduate with computer knowledge was preferred by them as a manager of the CIC.

8.3.5 Choice to take Initial Financial Support to Start CIC

Table 17: Respondents according to their expectation of provision of financial facility

N=100

Sr. No	*Institution*	*Number*	*Rank*
1	Government	2.00	I
2	Co-operative societies	0.92	III
3	By villagers' sharing	0.07	V
4	Voluntary donation	0.65	IV
5	Foreign relatives	1.45	II

With a view to knowing farmers' choice of source to take initial financial support to start CIC, information was collected and presented in Table 17. The first choice of the farmers to take initial financial support to start CIC was Government agencies followed by co-operative societies, villagers sharing, voluntary donation and foreign relatives.

8.3.6 Choice to Bear Expenses to Run CIC

The information regarding choice of the farmers to bear expenses to run CIC was also collected and presented in Table 18. It can be seen that respondents said that Government should bear expenses to run CIC, at the same time other preferences given by them to bear expenses to run CIC were village panchayat, co-operative dairy and collecting charges from users.

Table 18: The respondents as per the choice to bear expenses to run CIC

N=100

Sr. No.	*Responsible for expenditure*	*Number*	*Rank*
1	Government	1.60	I
2	Village Panchayat	1.02	II
3	Co-operative Dairy	0.92	III
4	By collecting charges from users	0.79	IV

8.3.7. Choice to have Minimum Computer sets at CIC

In present study respondents were also asked to give there expectation regarding number of internet connected sets of computers. The result says that majority of the farmers (78 per cent) suggested that more than six sets should be there at CIC while 22.00 per cent expected 5 to 6 sets of internet connected computer sets at CIC.

Table 19: The Respondents' choice to have minimum computer sets at CIC

N=100

Sr. No.	*Internet sets*	*Number*	*Per cent*
1	1 to 2	00	0.00
2	3 to 4	00	00.00
3	5 to 6	22	22.00
4	More than 6	78	78.00
	Total	100	100.00

8.3.8 Expectation about the form of Information on Internet and at ICT

The form of information available on farmers' related webs or sites and on the ICT plays important role in making farmers more convinced about any new things. Thus information was collected to know their preferences about it. It can be seen that information was expected by most of the farmers in Gujarati language that too in form of photographs and written form on agricultural related webs as well as on ICT. At the initial stage if it is difficult to provide information on most expected form, thus many of them were also expected information in audio-visual form, written form and audio form.

Table 20: Respondents choice on form of information on internet and ICT

N=100

Sr. No	*Type of information*	*Mean value*	*Rank*
1	In Gujarati language	1.98	I
2	Photographs with written form	1.94	II
3	Audio-visual form	1.91	III
4	Written information	1.42	IV
5	Audio form	1.20	V

The result indicates that farmers are ready to use Internet for agricultural information but on majority of the agricultural sites, information is available in English language so they are not in opposition to use it. Thus, all State Government should launch farmers' related sites in local languages. It is interesting to note that many State Governments have initiated their efforts in this direction. Gujarat Government had also started efforts to launch site for the farmers in Gujarati language. Anand Agricultural University is playing it key role in providing technical and information supports for this venture.

8.3.9 Purposes of to have CIC

There are many uses of Internet; person who uses it can explore it for his or her own areas of interest. Here farmers were asked to show their purposes of Internet use. The data presented in Table 21 reveals that the major as uses of internet expected by the respondents were to collect agricultural information, to collect information on government's programmes, for speedy communication, for exchange information, for entertainment, to collect information for their children's education, to contact foreign relatives, to contact relatives in India, to know information on agriculture of developed county and to know more about market prices.

Table 21: Respondents according to their purposes to have CIC

N=100

Sr. No	*Purpose*	*Mean value*	*Rank*
1	To collect agricultural information	1.71	I
2	To collect information on Government's Programmes	1.64	II
3	For speedy communication	1.61	III
4	For exchange information	1.21	IV
5	For entertainment	1.11	V
6	To collect information for their children's education	0.74	VI
7	To contact foreign relatives	0.67	VII
8	To contact relatives in India	0.64	VIII
9	To know information on agriculture of developed county	0.56	IX
10	To know more about market prices	0.51	X

Thus, from this result it can be said that farmers have various areas of interest to use Internet at CIC but the major five areas of their interest were to collect agricultural information, to collect information on government's programmes, for speedy communication, for exchange information and entertainment.

8.3.10 Expectation to Involve in the Process of Exploring Internet at CIC

With a view to finding their real interest to involve themselves in the process of exploring internet at CIC, their expectation was known. This information can give us picture that whether really they want to be self sufficient or dependent on others. The data presented in Table 22 indicate that majority (78.00 per cent) of the respondents were expecting to use internet by their own while, 22.00 per cent were wanted to use it with the help of others.

Table 22:Respondents' wish to involve in the process of exploring internet

N =100

Sr. No	*Way of internet use*	*Number*	*Per cent*
1	By own	78	78.00
2	With the help of others	22	22.00
Total		100	100

The result shows that farmers have real interest to be a part of Internet communication process thus, majority (78.00 per cent) of the respondents were expecting to use Internet by their own while.

8.3.11 Expected Frequency of the Farmers to use Internet at CIC

The expected frequency of the farmers to use Internet facility at CIC was measure. The result indicates that majority (63.00 per cent) of respondent were expecting to use internet at CIC daily or twice in a week, followed by 28 per cent whenever needed, 06.00 per cent once in a week and 3.00 per cent once in a fortnight. Thus it can be said that favorable expectation was observed among the farmers to use Internet at CIC.

Table 23: Respondents expected frequency to use internet at CIC

N=100

Sr. No	Expected frequency to use ICT	Number	Per cent
1	Daily	32	32.00
2	Twice in a week	31	31.00
2	Once in a week	06	6.00
3	Once on fortnight	03	3.00
4	Once in month	00	00.00
5	Whenever needed	28	28.00
	Total	100	100.00

12.3.12 Expected Agency to Receive Training to use CIC

When respondents were asked to show about their expectation regarding taking the training to use internet facility, cent per cent of them expressed their positive response to take proper training about the use of internet facility at available at CIC. Their expectations were also known in terms of agency of training. The data presented in table indicate that majority (70.00 per cent) of respondents expected training through government agency, followed by 13.00 per cent expected training by Agricultural University, 11.00 per cent respondents by Gram Panchayat. Only 4.00 and 2.00 percent of them expected such training by co-operative society and NGOs.

Table 24: Respondents expected agency to receive training to use CIC

N =100

Sr. No	Agency	Number	Per cent
1	Government	70	70.00
2	Co-operative society	04	04.00
3	NGOs	02	02.00
4	Gram Panchayat	11	11.00
5	Agricultural University	13	13.00
	Total	100	100.00

Thus, the data can be concluded that majority (70.00 per cent) of the respondents expected training of Internet through Government agencies.

9. SUMMARY AND CONCLUSION

From this study the following findings have been obtained.

1. Majority of the respondent farmers (64.00 per cent) had middle age and slightly less than half (45.00 per cent) of them had education up to high school or higher secondary level.
2. Majority of the farmers had agriculture + Animal husbandry as major sources of income (58.00 per cent), small size of landholding (60.00 per cent) and two or more than two miich animals (72.00 per cent), while (24.00 per cent) of the respondents were having tractors.
3. Majority of respondents had joint type of family (62.00 per cent) and membership in one organization (55.00 per cent), while 10.00 per cent of them had their relatives in abroad countries. The television was used by highest number of the respondents followed by radio, newspaper and other, respectively.
4. The farmers opinion towards Internet, illustrates their keen awareness for this particular technology. 69.00 per cent farmers have nodded their heads when they were asked whether Internet be a very useful mean to the farmers during present time, 10 per cent of them were partially agreed and remaining 21 per cent were still in confusion. Majority (71.00 per cent) of the farmers have realized that Internet is a rich source to collect worldwide information on agriculture and its allied fields. Larger mass (72 per cent) of the farmers have accepted that Internet is fastest way to exchange information in shortest time. As this technology is in its preliminary stage, it seems somewhat a costly affair for the farmers, but they are still hopeful. 41.00 per cent of the farmers have supported that Internet is the best mean to collect information on market prices of agricultural products. At the same time the most of the farmers (61.00 per cent) got disagreed with the statement that Internet is just a time pass activity. 47 per cent of the opinion also supported that

information available on Internet is easy to understand for the farmers. Most of them (56 per cent) said that development of Indian farmers is possible through Internet. Internet would be a blessing boon for the new generation hence the farmers are optimistic enough to let their children make positive use of Internet technology. So looking at the advantages of Internet Technology, it is obvious that they strongly believe that farmers should make a wide use of Internet. The power of a single click has amazed the world since the revolutionary Information Technology has arrived. It has enhanced the accessibility of the resources to one level up. Agriculture is also one of the beneficiaries. Today's farmers have positive personal perceptions concerning Internet technology.

5. The variables like Education, Land holding, Contact with NRI's, Experience of internet use and Mass media exposure were observed positively significant while, type of the family was observed negatively significant with the opinion of the farmers about the use of Internet for farming community. The independent variables like Age, Occupation, Animal wealth, Modern agricultural equipment, Type of family, Extension contact and Organization participation were observed non significant with the opinion of the farmers about the use of Internet for farming community.
6. Cent per cent respondent was expected the Community Internet Centre (CIC) facilities at village level.
7. Slightly less than cent per cent (95.00 per cent) of the respondents preferred panchayat office as the best place of CTC.
8. Majority of the farmers were expecting operator or guide (98.00 per cent), printer (72.00 per cent), separate cabin (68.00 per cent), downloaded information in printed form (63.00 per cent) and extra seating facilities (53.00 per cent) at CIC.
9. The result indicates that agriculture graduate with computer knowledge was preferred by nearly cent per cent (98.00 per cent) of the farmers as a manager of the centre, followed by any educated person of the village was preferred by 76.00 per cent, expert of computer by 57.00 per cent and primary

school teacher by 28.00 per cent of the farmers. Here farmers suggested multiple choices for the person as a manager of the CIC.

10. The first choice of the farmers to take initial financial support to start CIC was Government agencies followed by co-operative societies, villagers sharing, voluntary donation and foreign relatives.
11. The respondents said that Government should bear expenses to run CIC, at the same time other preferences given by them to bear expenses to run CIC were village panchayat, co-operative dairy and by collecting charges from users.
12. The result says that majority of the farmers (78 per cent) suggested that more than six sets should be there at CIC while 22.00 per cent expected 5 to 6 sets of Internet connected computer sets at CIC.
13. The information on CIC and on the farmers' related sites was expected by most of the farmers in Gujarati language that too in form of photographs and written form. At the initial stage if it is difficult to provide information on most expected form, many of them were also expected information in audio-visual form, written form and audio form.
14. The major purposes to have CIC explained by the respondents were to collect agricultural information, to collect information on government's programmes, to speedup communication, to exchange information, to have some entertainment, to collect information for their children's education, to contact foreign relatives, to contact relatives in India, to know information on agriculture of developed county and to know more about market prices.
15. Majority (78.00 per cent) of the respondents were expecting to use Internet by their own while, 22.00 per cent were wanted to use it with the help of others.
16. The expected frequency of the farmers to use Internet facility at CIC was measured. The result indicates that majority (63.00 per cent) of respondent were expecting to use internet at CIC daily or twice in a week, followed by 28 per cent whenever needed, 06.00 per cent once in a week and 3.00 per cent once in a fortnight. Thus it can be said

that favourable expectation was observed among the farmers to use Internet at CIC.

17. The cent per cent of the respondents expressed positive response to have proper training about the use of Internet facility at available at CIC. Out of which (70.00 per cent) of respondents expected to have training through government agency, followed by 13.00 per cent expected training from Agricultural University, 11.00 per cent respondents from Gram Panchayat. Only 4.00 and 2.00 per cent of them expected such training by co-operative society and NGOs.

10. LITERATURE CITED

Aditya, V. P. S. (2003). ICT for knowledge sharing and enabling supply chain for green produce and products, Paper presented at the National Workshop on "ICT for Agriculture and Rural Development" organized by Indian Association for Information Technology in Agriculture and Dihrubhai Ambani Institute for Information and Communication Technology (DA-IICT), Gandhinagar (Gujarat), India on 18th December 2003.

Anonymous (2001). INSTANT MESSAGING http://www.realtor.org/webintell.nsf/0/9d9b6f0d9a364b7886256aa900510d.

Anonymous (2002). Uses of Instructional Technology in Courses: A Survey of Student Views and Experiences, http://www.umass.edu/oapa /assessment/instruct_tech_in_courses.pdf.

Anonymous (2002)*. Use of information technology in agriculture, http://www.indiainfoline.com/bisc/itin.pdf.

Anonymous (2004). E-Commerce Sector Impact Study – Conclusions (2004) http:// strategis.ic.gc.ca/ epic / internet / inee-ef.nsf / vwGenerated Inter E/ee00464e.html.

Andrew Trotter (2002). Internet Access Has No Impact on Test Scores, Study Says,http://lsc-net.terc.edu/do.cfm/paper/8509/show/use_set-tech.

Barbara A. Walczak (1996) The Quality and Quantity of Elementary Students' Use of the Internet as a Resource in Science Problem Solving, http://essc.calumet.purdue.edu/Activity%20Science%20Research/chapter%20six.

Buddhadev, B. V. (2003). Crop Production Information System for farmers, Paper presented at the National Workshop on "ICT for Agriculture and Rural Development" organized by Indian Association for Information Technology in Agriculture and Dihrubhai Ambani Institute for Information and Communication Technology (DA-IICT), Gandhinagar (Gujarat), India on 18th December 2003.

Campbell J., Richard C. Sherman, Egon Kraan, & Zachary Birchmeier (2001) Internet Privacy Awareness and Concerns among College Students, http://www.users.muohio.edu/shermarc/aps01.htm

Chaudhary, S, Patel , A and Christian End, Egon K, Alison C, Jamonn C, Zachary B, Jaime K and Richard C. Sherman (2000). The Internet Gender Gap Among College Students, http://www. users.muohio. edu /shermarc/mpa100.htm.

Curtis J. Bonk (2002). Online Training in an Online World, HYPERLINK http://aftnn.org/blanket/site/http://www.usdla.org/html/ journal/MAR02_Issue/article02.html.

Erwin, P. (2001). Attitudes and persuasion. Hove: Psychology Press Limited. http://www.ied.edu.hk/has/webauth/4hkws.

Gee, D. B. (1990). The impact of students' preferred learning style variables in a distance education course: A case study, ERIC Document Reproduction Service No. ED 358836. http://www.ied. edu. hk/has/webauth/4hkws.

Goh, C. W (1997) Use and acceptance of Internet as a modern communication tool: a study in the city of Kuala [Lumpur]. A Baccalaureate Degree Project Paper. Department of Communication, Universiti Kebangsaan Malaysia. (Unpublished). Referred from the article on "Internet in Malaysia" by Rahmah Hashim and Arfah Yusof, http://www.interasia.org/malaysia/ hashim-yusof.html.

Kerlinger, F. N. (1976). Foundation of Behavioural Research, New Delhi. Surjeet Publication, pp. 198-204.

Kohang, A. A. (1989). A study of attitudes toward computers: Anxiety, confidence, liking, and perception of usefulness. Journal of Research on Computing in Education, 22(2), 137-150. http://www. ied.edu.hk /has/webauth/4hkws.

Lee, P. L. (1997). Penggunaan Internet di Lembah Kelang: satu kajian di Subang Jaya. [Use of Internet in the Klang Valley: a study in Subang Jaya]. A Baccalaureate Degree Project Paper. Department of Communication, Universiti Kebangsaan Malaysia. (Unpublished). Referred from the article on "Internet in Malaysia" by Rahmah Hashim and Arfah Yusof, http://www.interasia.org/malaysia / hashim-yusof.html.

Moni Madaswamy (2003). Digital Opportunities – A Powerful tool for fostering Agricultural Growth, Poverty Reduction and Sustainable Resource Use in India, * Theme Paper presented at the National Workshop on "ICT for Agriculture and Rural Development" organised by Indian Association for Information Technology in Agriculture and Dihrubhai Ambani Institute for Information and Communication Technology (DA-IICT), Gandhinagar (Gujarat), India on 18th December 2003.

Mayes, R. L. (1995). The application of a computer algebra system as a tool in college algebra. School Science and Mathematics, 95(2), 61-68. HYPERLINK http://www.ied.edu.hk/has/webauth/4hkws.

Pang, W. L. (1997). Penggunaan Internet di Lembah Kelang: satu kajian kes di Petaling Jaya. [Use of Internet in the Klang Valley: a case study in Petaling Jaya]. A Baccalaureate Degree Project Paper. Department of Communication, Universiti Kebangsaan Malaysia. (Unpublished). Referred from the article on " Internet in Malaysia" by Rahmah Hashim and Arfah Yusof, http://www.interasia.org/ malaysia/ hashim-yusof.html.

Patel, D.D and Patel, A. A. (2003). How ICT Tools can Help Support TOT in Agriculture, Paper presented at the National Workshop on "ICT for Agriculture and Rural Development" organized by Indian Association for Information Technology in Agriculture and Dihrubhai Ambani Institute for Information and Communication Technology (DA-IICT), Gandhinagar (Gujarat), India on 18th December 2003.

Pareek, U. and Trivedi, G. C. (1956). Factor analysis of Socio-economic status of farmers in India. Rural India, 30 : 311-321.

Purvi Mehta-Bhatt (2003). Information Technology in Agriculture: Reaching the Un-reached, Theme Paper presented at the National Workshop on "ICT for Agriculture and Rural Development" organized by Indian Association for Information Technology in Agriculture and Dihrubhai Ambani Institute for Information and Communication Technology (DA-IICT), Gandhinagar (Gujarat), India on 18th December 2003.

Rangarao, V. (2003). ICT in the empowerment of Indian farming community: the success story of EENADU group, Theme Paper presented at the National Workshop on "ICT for Agriculture and Rural Development" organized by Indian Association for Information Technology in Agriculture and Dihrubhai Ambani Institute for Information and Communication Technology (DA-IICT), Gandhinagar (Gujarat), India on 18th December 2003.

Rathod Virendra (2003). Application of ICT to Enhance Agriculture, Theme Paper presented at the National Workshop on "ICT for Agriculture and Rural Development" organized by Indian Association for Information Technology in Agriculture and Dihrubhai Ambani Institute for Information and Communication Technology (DA-IICT), Gandhinagar (Gujarat), India on 18th December 2003.

Richard C. Sherman (1997) Social Psychology in Cyberspace: Using the World Wide Web to Teach Applications of Social Principles to Everyday Life. HYPERLINK http://www.users.muohio.edu/shermarc/scip97.htm

Rothman, A. H. (2000). The impact of computer-based versus 'traditional' textbook science instruction on selected student learning outcomes. Dissertation Abstracts International, 69 (03), 938. UMI No. 9614309. SAS Institute Inc., SAS/STAT User's Guide, Version 6, Fourth Edition, Volume 2, Cary, NC SAS Institute Inc., 1989. 846 pp. HYPERLINK http://www.ied.edu.hk/has/webauth/4hkws.

Shaik, N. M, Jhamtani Anita, Rao, D (2003). An Analysis of Agricultural ICT Projects in India: Implications to the Agricultural Extension Systems, Paper presented at the National Workshop on "ICT for Agriculture and Rural Development" organized by Indian Association for Information Technology in Agriculture and Dihrubhai Ambani

Institute for Information and Communication Technology (DA-IICT), Gandhinagar (Gujarat), India on 18th December 2003.

Sharon Gudin (2003). U.S. Companies Best Use IT for Competitive Advantage httpitmanagement.earthweb.com/itres/article.php/1566951

Sontakki, B.S. (2003). Information Technology in Agricultural Research and Education Management: Role of NAARM, Paper presented at the National Workshop on "ICT for Agriculture and Rural Development" organized by Indian Association for Information Technology in Agriculture and Dihrubhai Ambani Institute for Information and Communication Technology (DA-IICT), Gandhinagar (Gujarat), India on 18th December 2003.

Surya, R and Amardeep (2003). New ICT for farmers: Opportunities and Challenges, Information Technology in Agricultural Research and Education Management: Role of NAARM, Paper presented at the National Workshop on "ICT for Agriculture and Rural Development" organized by Indian Association for Information Technology in Agriculture and Dihrubhai Ambani Institute for Information and Communication Technology (DA-IICT), Gandhinagar (Gujarat), India on 18th December 2003.

Sushma Panigrahy (2003). Remote Sensing and GIS for Agriculture and Rural Development, Paper presented at the National Workshop on "ICT for Agriculture and Rural Development" organized by Indian Association for Information Technology in Agriculture and Dihrubhai Ambani Institute for Information and Communication Technology (DA-IICT), Gandhinagar (Gujarat), India on 18th December 2003.

Tang, K. Y. (1998). Kajian urusniaga elektronik melalui Internet di Universiti Malaya. [A study of electronic commerce via Internet in University Malaya]. A Baccalaureate Degree Project Paper. Department of Communication, Universiti Kebangsaan Malaysia. (Unpublished). Referred from the article on " Internet in Malaysia" by Rahmah Hashim and Arfah Yusof, http://www.interasia.org/malaysia/ hashim-yusof.html.

Teresa Y. N, Jessame F, Michael Romary, Simmona Simmons-Hodo and Katy Sullivan (2003) Literacy Survey – 2003, Executive Summary http://aok.lib.umbc.edu/reference/informationliteracy/ESinfolit2003.pdf

Vijayaletchumy (1998). Pengetahuan pelajar mengenai teknologi maklumat: satu kajian di Jabatan Komunikasi, UKM. [Students' knowledge about information technology: a study at the Department of Communication, UKM. A Baccalaureate Degree Project Paper. Department of Communication, Universiti Kebangsaan Malaysia. (Unpublished). Referred from the article on " Internet in Malaysia" by Rahmah Hashim and Arfah Yusof, http://www.interasia.org/ malaysia/hashim-yusof.html.

APPEDICES

Appendix – I

Vasad Village at Press due to CIC

Vasad gram panchayat goes online
SACHIN SHARMA

TIMES NEWS NETWORK [SUNDAY, FEBRUARY 17, 2002 12:17:46 AM]

http://www1.timesofindia.indiatimes.com/articleshow/1201234.cms

Vadodara over 500 NRI families from Vasad will now not miss their hometown, for it just a click away. The Vasad gram panchayat has gone online with its website www.vasad.com to keep them informed about the developments back home and take care of their problems pertaining to land and other minor issues. This village, known for its 'Tuvar dal', has seemingly decided to become it savvy with several programmes in the pipeline for the purpose. The main stress, however, remains Computerisation of the gram panchayat. With land sharks eyeing the first chance to grab property, which has been left unattended, security of their holding has always bothered NRIs. Besides, there are several of them whose kids have not seen the land where they belong to and wonder where their forefathers hailed from. The website launched last year, now plans to include land records and information regarding birth as well as death of village residents. "It always came to our notice that those living abroad were always worried about their property and the chances of manipulation of land records. The website will take care of their worries. They can see relevant details regarding land holding online," said sarpanch of the village Dinesh Patel. Mr. Patel said the new system would make it possible for people to enter their names and survey number of their land holding as input in the site. Following this they will be able to see the status of the ownership of their property, the tax due on it and other relevant details. Similarly, by simply entering the full name of a person, the information regarding his birth and death can be accessed. The person can also obtain a print of the certificate. "Legally, this information can be used as a temporary proof, a relevant certificate can be issued by the panchayat later," Patel says. The functionaries vouch for the security of the system they have designed. Land records and birth-death details can be changed only using a security password. Only three persons in the panchayat the sarpanch, secretary and computer operator have access to it. Panchayat functionaries point out that as many as 80 per cent of the persons residing in the village have relatives abroad. "they are always curious to know what is happening here. Besides, we too feel that they must be informed about the developments here. Today, we boast of education from primary school to graduation in engineering. There are several other new things they might not know about," says a panchayat employee. The website has been a result of the efforts of the panchayat along with technical inputs from students of the Sardar Vallabhbhai Patel institute of technology. the students designed the website free of cost for the panchayat. "We had absolutely no knowledge of computers when we began the work. The only idea was to take the panchayat into the information technology age. It was the students who helped us out with it. Even I could not operate a computer then, now I use it and surf the Internet as well," Patel says. The next step to be taken by the panchayat is the computerization of rationing shops. First to be computerized will be the biggest co-operative shop in the village. "The village has about 3,000 ration card holders. Out of these about 1,500 are registered with only one shop. We plan to enter the details of all these cards and record transactions made on each card. Besides, we will also provide bills from the computer print out," said Patel. Computers for the purpose have already been installed and will begin functioning soon. "We shall not only have details of transaction for a month or a year on every card, but also the details of population in the village. The process will thus solve two purposes at the same time," said Patel. The panchayat is now eyeing the award for the best village panchayat given by the state government. "With all the modernization and welfare work we are confident of bagging it," Patel says. Sarpanch Dinesh Patel is working on the gram panchayat website.

Appendix - II

Profile survey

(1) **Farmer's Name :** ________________

Village : __________ Taluku : __________ District : __________

(2) Age : ________ years

(3) **Education :** Illiterate ☐ Primary ☐ Secondary ☐ Higher secondary ☐ Graduate ☐

(4) **Profession :** Agriculture ☐ Dairy ☐ Service Others ☐

(5) **Family :** Joint ☐ Nuclear ☐

(6) **Caste :** General ☐ OBC ☐ ST ☐ SC ☐

(7) **Land holding :** Irrigated ______ acre, Unirrigated ______ acre

Total : ______ acre

(8) **Milking Animal :** Cattles ___, Buffaloes ___, Goat ___, Total ___

(9) **Modern agriculture equipments :** Tractor ☐ Thresher ☐

Power trillor ☐ Tractor mounted sprayer and duster ☐

(10) **Is your relatives stay in Abroad ?** Yes ☐ No ☐

(A) If yes then, contry : __________

(B) Relation : ____________ (C) Since ________ years

(11) **Through which medium you contact your relatives ?**

(A) Post ☐ (B) Telephone ☐ (C) Internet ☐

(12) **From where you are getting mass information regarding agriculture?**

Sr. No.	*Representative*	*Once in a week*	*Monthly*	*More than month*	*Never*
1	Village level worker				
2	Extension officer				
3	Scientist				
4	Progressive farmers				
5	Friends				
6	Neighbour				
7	Religious groups				

(13) Which media is used for mass contact ?

Sr. No.	*Media*	*Everyday*	*Weekly*	*Monthly*	*Never*
1	Television				
2	Newspaper				
3	Internet				
4	Radio				
5	Others literature				

(14) Membership in a social organization ?

Sr. No.	*Organization*	*Member*	*Position*
1	Farmers co-operatives		
2	Gram Panchayat		
3	Taluku Panchayat		
4	District Panchayat		
5	Milk co-operative		
6	Youth club		
7	Others		

(15) Knowledge test :

(i) Do you know internet ? Yes ☐ No ☐

(ii) If yes, how long __________

(iii) How you get an information from website ?

(A) By writing ☐ (B) By printer copy ☐

(iv) Charge of internet per hour __________ Rs.

(v) Which search engine is used for searching information ?

Yahoo ☐ Google ☐ India Times ☐ Rediff ☐ Ask ☐ Other ☐

(vi) Which website is used for agriculture information ?

(1) ______ (2) ______ (3) ______ (4) ______

(16) Opinions

(i) Internet is most useful for farmers at present.

Yes ☐ No ☐

(ii) Internet is a bank for agriculture and its related information

Yes ☐ No ☐

(iii) Quick exchange of information is possible through Internet.

Yes ☐ No ☐

(iv) At present internet is economically most feasible for farmers

Yes ☐ No ☐

(v) Internet is nothing but media for time pass

Yes ☐ No ☐

(vii) Information on internet is easily understandable.

Yes ☐ No ☐

(viii) The development of farmers will be possible through Internet.

Yes ☐ No ☐

(ix) I wish to educate my children for Internet use.

Yes ☐ No ☐

(x) I wish to advise other farmers to use the internet

Yes ☐ No ☐

(17) Expectations

(i) Do you want create group internet facility in your village ?

Yes ☐ No ☐

(ii) If yes, which place is ideal for this ?

(A) Gram Panchayat ☐ (B) Primary school ☐

(C) High school ☐ (D) Milk co-operative ☐

(iii) Which facilities you are expecting at group internet centre ?

(A) Operator ☐ (B) Facility to get information from internet ☐

(C) Good seating ☐ (D) Separate cabin ☐

(iv) Which person you are expecting to operator the group internet centre ?

(A) Educated person from village ☐

(B) Agriculture graduate ☐

(C) A person from any field ☐

(D) Computer expert ☐

(E) Person from village or outside the village ☐

(v) Who will bare the running expenditure of the group internet centre ?

Sr. No.	*Name of organization*	*Contribution*		
1	Milk co-operatives	Highest	Medium	Negligible
2	Gram Panchayat			
3	Social organization			
4	Farmers co-operative			
5	Farmers himself			

(vi) Who will provide the financial assistance for group internet connection at village level ?

Sr. No.	*Name of organization*	*Contribution*		
1	Government	Highest	Medium	Negligible
2	NGO			
3	Co-operatives			
4	NRI friends			
5	People participation			

(vii) How many computer set is required to fulfill Internet facilities at your village ?

(A) 1 to 2 ☐ (B) 3 to 4 ☐

(C) 5 to 6 ☐ (D) Less than six ☐

(viii) Information on internet in which format ?

Sr. No.	*From*	*High*	*Medium*	*Low*
1	Photographs			
2	Gujarati language			
3	Audiable information			
4	Written from			

(ix) For which purpose you wish to use internet ?

(A) To get information (B) Entertainment

(C) Exchange of information (D) Chatting with relatives

(x) Way of Internet use

(A) Self (B) With help of others

(xi) Frequency of Internet use

Sr. No.	*Time*	
1	Daily	
2	Twice in a week	
3	Once in a fortnight	
4	As per need	
5	Never	

(xii) If you wish Internet trainees for farmers

(A) Yes ☐ (B) No ☐

(xiii) If yes who provide

(A) Government (B) Co-operative society

(C) Voluntary Institution (D) Gram Panchayat

(E) Agriculture University

2 Information Needs of the Rice Growers in Gujarat

ABSTRACT

1. From this research it can be concluded that majority of the rice growers had middle age (52.00 per cent), up to secondary to higher secondary level of education (56.00 per cent), used fertilizer depot, neighbors an television as sources of information, low to medium level of social participation (74.00 per cent), medium level of extension contact (52.00 per cent), big size of land holding (53.00 per cent), 65 per cent of their total land holding under rice cultivation (64.00 per cent), 65 per cent of their total land under irrigation (66.00 per cent) and 50,000 to 10,000 rupees of annual income (52.00 per cent), while slightly more than two fifth of the rice growers had medium level of cosmopoliteness (42.00 per cent), economic motivation (41.00 per cent) and market orientation (45.00 per cent).
2. Information needs of the rice growers: Plant protection measures, Marketing, Schedule of water supply by Canal, Fertilizer management, Water management, Preparation of Seedlings, Variety, Land preparation and sowing, Supportive facts, Harvesting and post harvesting technology and Weed management.

   - Major information needs of rice growers about improved verities were sources of seeds, suitable high yielding variety for the area and rate of seeds.
   - The Information about water to be supplied in channel before sowing time and advance information about time and date of supply of water in canal for complete crop period were the expected information needs of the rice growers relate for the canal irrigation mangers.
   - How to select site for raising seedlings, Method of preparing bed for nursery, Plant protection in nursery management, Nutrient management in nursery, Irrigation management in nursery and Proper age to select seedlings for transplanting were the major information needs realized by the rice growers for nursery management in rice cultivation.

- The major areas information needs expressed by the rice growers about sowing of rice were land preparation, Soil treatment methods. Place of availability of soil treatment inputs, Seed rate, Price of soil treatment inputs, Sowing time, Depth of sowing, Method of sowing, Spacing, Seed treatment inputs and Gap filling.
- The major area of information needs related to fertilizer management were Price of fertilizers, Stock of fertilizers, Place of availability of fertilizers, Name of highly advantageous chemical fertilizers for rice, Method and time of fertilizer application, Nutrient requirements of plant, Calculating the doze of chemical fertilizer, Deficiency symptoms of major plant nutrients, Bio-fertilizers and procedure of making organic fertilizers.
- Some of the smallest amounts of information needs expressed by rice growers for weed management were Chemical weed control, Price of weedicides, Place of availability of weedicides, Trade name of weedicides, Stock of weedicides and Hand weeding.
- The rice growers expressed that Schedule for irrigation, Critical Stages of Irrigation, How to save crop during shortage of water, Fertilizer management during irrigation and Method of irrigation were key areas of their information needs regarding irrigation management in rice crop.
- The major rice information needs about plant protection measures expressed by rice growers were identification, nature of damage and control measures for insects/pests as well as diseases of rice crop; price of insecticides/pesticides; integrated pest management and method of preparing solutions of insecticides/pesticides.
- How to store rice production, Care after harvesting at farm level and Care during harvesting were major information needs of rice growers about harvesting and post harvest technology.
- Rice growers articulated high need for information on market price followed by quality parameters that affects price and time of market inflow.
- The rice growers expressed their needs for information on supportive facts like subsidies, insurance in rice crop and government policies related to rice cultivation.

3. The characteristics like Use of information sources, Extension participation, Land under rice cultivation and Cosmopoliteness were significantly related with their information needs for rice cultivation. In other words rice growers who had better contact with sources of information, extension personals and more area under rice cultivation as well as high level of cosmopoliteness realized more information on various aspects of rice cultivation.

1. INTRODUCTION

Agriculture is the strength of Indians economy, providing direct employment to about 70 per cent of working people in the country.

It forms the basis of many premier industries of India, including the textile, jute, and sugar industries. Agriculture contributes about 31 per cent to GDP; about 25 per cent of India's exports are agricultural products.

Rice is the staple food of 65 per cent of the total population in India. It constitutes about 52 per cent of the total food grain production and 55 per cent of total cereal production. In our country both food and nonfood crops are grown. Food grains consist of cereals such as rice, wheat, jowar, bajra, and maize as well as pulses. Food crops grow on nearly 70 per cent of the gross sown area.

India became self sufficient in rice in 1977. That was achieved through a combination of increasing the area under cultivation and increasing cropping intensity. With the adoption of modern varieties in 1966, an average annual increase of 2 per cent in rice yield has been attained. About 55 per cent of the rice area was planted to modern varieties in 1985. Rough rice production has exceeded 100 million tones annually since 1988; total production in 1994 was almost 120 million tones, with an average yield of 2.6 tons/ha. India regularly exports a small amount of high quality basmati (aromatic) rice. In 1995, India exported nearly 4.2 million tons of rice, in response to the large increase in demand in the world market.

Increasing productivity (i.e., raising yields) is the vehicle for development of the rice sector. Rice production can be increased either by increasing the area under rice cultivation or by increasing the productivity of current cultivation. Given the pressure on agricultural land and the competition from other, more lucrative crops, it may be difficult to significantly increase the land under rice cultivation. In fact, this has actually decreased in the last seven years. The only solution, therefore, is to increase the productivity of the area currently under cultivation. This will lead not only to increased production but also to more efficient production. Three main factors affect rice productivity: irrigation; research and development and its extension to the farmer; and the availability of credit.

Research and development are crucial to create and deliver a package of technology suitable to the farmer's conditions. Particularly adaptive research is necessary to develop new technologies such as new high-yielding varieties, suited to particular conditions. For example, farmers in an area infested by a particular kind of pest should be able to use seed resistant to that pest. Similarly, farmers in drought-prone regions or in water-logged areas should use seed varieties that can withstand these conditions. The State should encourage the use of biotechnology to spur research of this kind. Furthermore, to be truly useful, the research carried out needs to be extended to the farmer. Effective extension services are extremely critical for improving rice yields. A large part of the credit for China's high rice yields goes to its well-developed extension system. Extension services include not only instructing farmers on the correct cropping 'package' they should use, but also educating them on good farm and crop management practices, e.g., correct plant population density, appropriate post harvesting technologies, and correct water and soil management practices.

Information is considered as a critical input in decision-making if research and extension efforts. Information is known as power, those who own it. In contrast, lack of information causes uncertainty about the possible impact of decisions on results. Lack of accurate and timely information is one of the important factors for the inefficiency and imperfections observed in the agricultural marketing system in India. Now, 'Information' has started gaining importance. Besides the development of the theory of information, the need to have information systems to facilitate sound decisions in agriculture was greatly felt. There is a need to have an integrated system providing the required information to the various decision-makers at the right time and place.

Agriculture sector would demand for quickly approaching quite a larger area with a variety of information right from inputs, production and market intelligence. It is not feasible to reach in person regularly to every farmer to convey high tech farming techniques. Hence, the future extension strategy would concentrate on distant extension education by way of newspapers, radio, TV or electronic media. Further, all these media would require to carry information as per the demand of the producers.

The preparation of good content of information of rice farming is possible based on the real information needs of the farmers. The content based on actual needs of the users will create interest among then to apply it in practice. With a view to supporting larger group of rice growers with agricultural information in future, it seems worthwhile to determine the information needs of the rice growers.

Explanation of Problem:

The research generates technologies that are carried by transfer of technology agents and thereby are used by farmers and ultimate users. It is claimed that quite a large number of technologies have been generated by the agricultural scientists which is very much useful to the farmers. Simultaneously, it is obvious that except few technologies including mainly some good crop cultivars, not many technologies have been either reached to the farms on a massive scale or the farmers really realized its contribution to increase production in short run or to increase sustainability in long run. The reasons may be many. Over and above the diverse situations prevailing in the fields as well as amongst the farmers, the capability of the farmers to purchase the inputs and non-availability of trustworthy inputs in time, high cost of purchased

inputs and un-assured outcome from the farming and above all, the technologies that can be used by them are either not known to them at the time they need it or they lack proper understanding of the technologies so as to try it to their farm in a right manner. It is needless to state that beyond producing appropriate technologies, it is equally important to take these technologies to all concerned.

This requires information management and its dissemination, training for knowledge and skill up gradation, providing platform for discussing, debating and making trial of the ideas and issues, etc. In addition to this, mass dissemination of information is also more important in view of its larger area coverage.

Acquisition of information has always been regarded as a factor playing an important role in molding human behaviour leading to decision for adopting of innovation. Identifying information needs of the users can become solid basis for developing meaningful information warehouse.

In Gujarat, number of new technologies for rice cultivation is generated. Till these technologies are taken to the farmers at the time they need it, the adoption rate would be low. Thus, for increased use of these technologies by the farmers, a right type of technology blended with right time would be required to be followed. Not only that the Khambhat taluku of Anand district carries significant area and production of rice in the district. There are more possibilities to harvest more. This is possible only if the farmers are provided with the information that is required by them, at the right time.

Keeping in view the significance of the information needs of the rice growers and as no study of this kind to assess the information needs of rice growers have been conducted in past, the present study was planned. The results of the study will form a basis for all those who are preparing contents for information to be transmitted to the rice growers for their use.

Objectives of the Study

The present study was undertaken with following objectives:

1. To study the profile of the rice growing farmers.
2. To ascertain the information needs of the rice growers.
3. To study the relationship between selected personal, social communicational, economical and psychological characteristics of rice growers and their information needs.
4. To suggest a model of technology transfer to cater the information needs of the rice growers.

Review of Literature

Reviews constitute an important source of information. It helps in clearing some concepts and improving focus help in avoidance of unnecessary duplication in research. In accordance with objectives of study the literature has been reviewed and presented as under

Patel (2004) reported that majority of the cotton growers expressed their needs for information about various aspects of marketing, water management, plant protection measures, fertilizer management and variety. Independent variables like use of information sources, extension participation, land under cotton cultivation and cosmopoliteness had significant correlation with information needs of the cotton growers.

Saini (1981) found that information need hierarchy were seed production, improved varieties, haymaking and fodder mixture and climatic requirements.

Stevens and Koch (1990) indicated that full and part time peach farmers in Silverton Extension ward did not differ significantly in their information needs.

Singh and Aggarwal (1993) found that information on harmful effects of drugs and smokeless chullah was most needed by Punjabi farmwomen. Fuel saving devices, time and energy saving devices and care and repair of household equipments were placed at third, fourth and fifth positions respectively.

Hanuman *et al.* (1997) reported that the tribal farmers had highest information needs about soil relation (78.00 MPS) where as the non-tribal farmers had the highest information needs about fertilizer application (74.17 MPS). It was also concluded that the tribal farmers had low information needs about use of biogas (30.34MPS) while the non-tribal farmers had poor information needs about fruits and vegetables (37.17MPS).

Sankar Rao and Reddy (1997) observed in their study on information needs of tribal farmers towards mango production technology that majority of the respondents (64.00 per cent) belonged to medium information need group followed by high (22.67 per cent) and low (13.33 per cent) information need group.

Pawar *et al.* (2001) observed that field crops and nursery management were considered as most important topics for information, followed by horticulture and farm development schemes.

Meera *et al.* (2003) found under Gyandoot Project (AICT Project in M.P. State) that majority farmers perceived market information, facilitation of land record, question-answer services, information on Rural Development Programs and weather forecasting as the most important and essential information needs.

Singh *et al.* (1991) revealed that knowledge gap or citrus growers is negatively and significantly correlated with their extension contact.

Talati (1991) reported that education had significantly exerted their influence on training needs of banana growers of Kheda district. He also found that there was no significant influence of irrigation facilities on training needs to collect information on banana cultivation of banana growers of Kheda district with respect to improved banana cultivation technology.

Dolli and Sudaraswamy (1994) revealed from their study on influence of socio-economic factors on technological information adoption gap in cultivation of pulse crop that there was negative and significant correlation between technological gap and extension contact.

Raut *et al.* (1995) observed in their study on influence of the characteristics of young farmers on their training needs to collection information about cultivation of fruit crops that education was significantly associated with training needs of the young farmers growing fruit crops.

Satpathy (1995) found that there was no significant association between age of the potato growers of Kheda district and their training needs on the information of potato cultivation. He also observed negative and significant correlation between education of the potato growers and their training needs on the information of potato cultivation. At the same time revealed that there was negative and significant correlation of land holding and social participation with training needs on the information of potato cultivation of potato growers.

Vaghela (1996) revealed that age of the banana growers of Anand taluku of Kheda district had established positive and significant correlation (0.18699) with their training need to collect information on modern practices of banana. He further concluded that economic motivation of banana growers of Anand taluka of Kheda district had negative and significant correlation (-0.65954) with the training need to collect information on modern practices of banana.

Anonymous (2004) Rice is the staple food for 65% of the total population in India. The Indian population was about 1 billion people in 2000 and is still growing at a high rate (1.7% per year). Although the country exports several varieties of rice, many scientists have expressed concern that current Indian rice production techniques cannot sustain the growing domestic population. India has a large number of rice dishes and many of them are very simple to prepare. Indian pilaf rice is very flavorful and fluffy when cooked with Basmati rice.

Anonymous (2004) About 600 improved varieties of indica rice have been released for cultivation since 1965, but Basmati rice is still planted over large areas. Rice-based production systems provide the main income and employment for more than 50 million households.

Anonymous (2003) Rice is the main source of food for about half the world's population. Although it is cultivated world-wide, 91 per cent is grown in Asia, which produces over 535 million metric tons of rice per year. On average, in the West, we consume 1.8 kg/4 lbs of rice per year per person, compared to 150 kg/330 lbs of rice per year per person in Asia.

2. RESEARCH METHODOLOGY:

2.1 Location of Research Area

In view of limited resources available with the investigator, the present investigation was confined to Anand district of Gujarat state. The district is located in middle of the Gujarat from the total cultivable area, large area is covered under irrigation. The principle crops of the district are tobacco, paddy, pearl millet, wheat and vegetable crops. Anand district comprises of eight talukus. Khambhat is the major rice growing taluku of the district. Thus study was undertaken in this taluku of the district. This taluku was selected purposely because area under rice cultivation is the highest among all the eight talukus of district.

2.2 Sample Size

In present investigation, simple random sampling technique was employed. First rice growing six villages viz; Gudel, Galiyana, Naviakhol, Rohini, Tamsa and Golana were selected to select respondents. The lists of rice growers were obtained for each of the selected villages from the gram panchayat office. It was planned to select almost 15 to 20 respondents from each of the selected villages randomly. Ultimately, a random sample of total of 100 respondents was selected for the study.

2.3 Research Design

As the study was concerned with understanding the information needs of rice growers, ex-post-facto research design was used for this study. This design was used because the researcher did not have any control on the independent variables of the selected rice growers.

Ex-post-facto research design is systematic experimental investigation in which control on independent variable is very weak and in this study all most no control was possible. Kerlinger (1976) stated that ex-post-facto research design is worthy to apply when the independent variables have already acted up on.

2.4 Derivation of Hypothesis

On the basis of objective of the study, the following null hypothesis was formulated. H_0 - there will not be any relationship between selected characteristics of rice growers and their information needs.

2.5 Limitations of the Study

On account of limited time and resources available with the investigator, the study was conducted with the following limitations:

1. The area of the study comprised to a few villages of Khambhat taluku of Anand district only. As such, the findings may be applicable only to the areas or places having similar agro-climatic conditions. A careful endeavor can help applying the findings of this study to other areas of the state.
2. The study was limited to 100 respondents.
3. The findings of the study were based upon the expression and responses of the rice growing farmers and its perception by the researcher. The validity of findings will therefore be limited to the extent the respondents are objective in their responses and the extent to which the researcher understood the responses correctly.

2.6 Method of Data Collection

An interview schedule was specially prepared and used as a tool for collection of requisite information. In all, 100 respondents were interviewed personally, with the help of the interviewed schedule. To avail correct and complete required information from respondents; an interview schedule was specifically structured keeping in view the objectives of the study. In formulating the questions and statements for schedule, the investigator sought the technical guidance from scientists/experts of the field and available literature and made it precise, clear and meaningful. The Data were collected by personal interview of the selected respondents. Every possible care was taken to seek an unbiased opinion of the respondents.

2.7 Selection of the Variables

The variables under study were selected on the basis of extensive review of literature related to the subject and consultation with experts and finally the variables that are found to be most relevant to the present study are selected as under.

2.7.1 Dependent Variables

1. Information needs of rice growers

2.7.2 Independent Variables

I. Personal characteristics
   (1) Age
   (2) Education
II. Social-communicational characteristics
   (1) Use of information sources
   (2) Social participation
   (3) Extension participation
III. Economical characteristics
   (1) Size of land holding
   (2) Land under rice cultivation
   (3) Irrigated area to total land
   (4) Annual income
IV. Psychological characteristics
   (1) Cosmopoliteness
   (2) Economic motivation
   (3) Market orientation

2.8 Measurement of Variables

2.8.1 Dependent Variable

2.8.1.1 Ascertaining Information Needs

The information needs of the rice growers were assessed for the areas right from the requirements for selection of crop and variety, through the availability of inputs, production technologies, harvesting and packaging; till farm level processing and marketing of the produce. Various related items were selected for these areas by reviewing the literature and a final shape was given after getting the opinion of the experts. Thus, the information needs of rice grower farmers about various aspects of rice cultivation were ascertained. The extent of information needs of the rice growers was availed on a three-point continuum ranging from most needed, needed and not needed. The three categories were assigned with 2 score, 1 score and 0 score respectively. The information needs were worked out for each of the major areas considering the total score for information need acquired by the respondents. On the basis of the mean score, the ranks to the major areas of information needs were assigned.

2.8.2 Measurement of Independent Variables

2.8.2.1 Age

Age refers to the total number of years completed by the respondents at the time of interview. The respondents were asked for their age in terms of completed years. The age in years as such was considered as the age score for each respondent. They were grouped in to three categories viz; Young (Up to 30 years), Middle (31 to 50 years) and Old aged (Above 50 years)

2.8.2.2 Education

It was operationalized as the level of formal schooling completed by the respondents. For each category, the scoring was assigned as per socio-economic scale of Pareek and Trivedi (1965), with due modifications. The one score for each year of the formal education was assigned to work out strength of this variable. They were classified in to five groups viz; Illiterate, can read and write, Primary education (1st to 7th std.), Secondary / Higher Secondary education (8th to 12 std.) and College and above education.

2.8.2.3 Use of Information Sources

Communication sources were conceptualizes as the sources through which Gujarat farmers get information about improved agricultural technologies, especially in regard to rice cultivation. To measure the degree of utilization of the sources, each respondent was asked to indicate; as to how often he accessed information about rice cultivation from each of the sources. Score of 2, 1 and 0 was assigned for each of the respondent, for response in terms of often, some times and never respectively. The mean score was worked for each source to know degree of its use.

2.8.2.4 Social Participation

Social participation in the present study was operationalized as the degree to which an individual was associated with different social organizations either as a member or its office bearer. One score was given for having membership in each organization and 2 score for holding position in each organization; whereas, 0 score was given for non-membership in any organization. The total score obtained by each respondent was considered for studying the variable. On the basis of total score, the respondents were grouped into three categories viz. low (below mean – 0.5 S. D.), medium (in between mean + 0.5 S. D.) and high (above mean + 0.5 S. D.).

2.8.2.5 Extension Participation

Extension participation is defined as the degree to which an individual participates in various non-formal extension activities, with a view to obtain new information, knowledge and skills related to agriculture. One score was given for participating in each of the extension activity. The respondents were grouped into three categories viz. low (below mean – 0.5 S. D.), medium (in between mean + 0.5 S. D.) and high (above mean + 0.5 S. D.).

2.8.2.6 Size of Land Holding

Land holding is an important factor that determines the economic status and potentiality of an individual respondent. It refers to the number of hectares of land owned and cultivated by the respondents. Land possessed by an individual including irrigated and un-irrigated land was considered as total size of land holding. After getting the total land holding of individual respondents, they were classified in to four categories viz; Marginal farmers (Up to 1.00 hectare), Small farmers (1.01 to 2.00 hectares), Medium farmers (2.01 to 4.00 hectares) and Big farmers (Above 4.00 hectares).

2.8.2.7 Land Under Rice Cultivation

The area under rice cultivation in hectares possessed by the respondents was considered for measuring this variable. Expert opinion was obtained for deciding range for classification of the respondents. Considering the opinion received from 25 experts, the respondents were grouped in to three categories; land under rice crop Up to 35.00 per cent, 35.01 to 65.00 per cent and Above 65.00 per cent of total land.

2.8.2.8 Irrigated Area to Total Land Holding

The proportion of irrigated area to total land holding of the respondents in hectares was considered for measuring this variable. Expert opinion was obtained for deciding range for classification of the respondents. Considering the opinion received from 25 experts, the respondents were grouped viz; land under irrigation Up to 35.00 per cent, 35.01 to 65.00 per cent and Above 65.00 per cent of total land.

2.8.2.9 Annual Income

This refers to the total annual earnings of the respondents through all sources. It was measured in terms of range and grouped into three categories viz; Low (Up to Rs. 50,000/- Medium (Rs. 50,001 to Rs.100,000/) and High (Above Rs. 100,000/-).

2.8.2.10 Cosmopoliteness

Cosmopoliteness is the tendency of an individual to be in contact with outside his own community, based on the belief that individuals all the needs cannot be satisfied within his own community. This was measured with the help of scale developed by Singh (1973). The total score for individual respondents was obtained by adding all the scores of the statements as suggested in scale. The sum of score attained for classifying into three groups viz; Low level of cosmopoliteness (Below mean - 0.5 SD), Medium level of cosmopoliteness (Mean + 0.5 SD) and High level of cosmopoliteness (Above mean + 0.5 SD).

2.8.2.11 Economic Motivation

This was measured with the help of economic motivation scale worked out by Supe (1969). Economic motivation score of an individual respondent was the sum of the scores of all the statements included in the scale. Categories formed on the basis of mean and standard deviation; viz; Low level (Below mean –0.5 SD), Medium (Mean + 0.5 SD) and High level of economic motivation (Above mean + 0.5 SD).

2.8.2.12 Market Orientation

The scale used to measure this variable is based on market orientation scale developed by Samantha (1977). Total score was obtained by summing all the responses. High score indicated high degree of market orientation in this scale. Conceptually, the respondents were categorized into three groups on the basis of mean and standard deviation; Low level of market orientation (Below mean –0.5 SD), Medium level of market orientation (Mean + 0.5 SD) and High level of market orientation (Above mean + 0.5 SD).

2.9 Statistical framework for the Analysis of Data

The following statistical tools were used for interpreting the data.

2.9.1 Frequency and Percentage

Simple comparisons were made on the basis of frequency and percentage.

2.9.2 Mean

The mean was obtained by the total number of score divided by total number of trainees.

2.9.2.1 Coefficient of Correlation

Coefficient of Correlation was computed to find out the relationship between each of the independent variable and the dependent variable by employing following formula.

$$r = \frac{N \Sigma XY - \Sigma X \Sigma Y}{\sqrt{\Sigma N \Sigma X^2 - (\Sigma X)^2] [\Sigma N(\Sigma Y) - (\Sigma Y)^2]}}$$

Where r = correlation coefficient, X = Independent variable, Y = Dependent variable and N = Total number of respondents.

3. RESULTS AND DISCUSSION

The data pertaining to the present investigation were collected, classified, tabulated, analyzed and presented in succeeding pages under following heads:

3.1. Characteristics of the Rice Growers

Information assumes value in a decision-making context. Information needs may depend on personal traits of rice growers. Personal traits in this study, includes personal characteristics, social-communicational characteristics, economical characteristics and psychological characteristics of rice growers. An attempt has been made to analyze some of the personal characteristics like age and education; social communicational characteristics like use of information sources, social participation and extension participation; economical characteristics like size of land holding, land under rice cultivation, irrigated area to total land and annual income; psychological characteristics like cosmopoliteness, economic motivation and market orientation. The results obtained in this investigation are presented as under:

3.1.1 Age

Age is an important factor, which might have some influence on the behaviour pattern of the rice growers. Generally, it is believed that young farmers are active and dynamic. The data presented in the Table 3 reflect that more than half of the respondents (52.00 per cent) were in the middle age group followed by old age group (36.00 per cent) and young age group (12.00 per cent).

Table 1: The respondents according to the age

N = 100

S.No	*Age Group*	*Number*	*Per cent*
1.	Young	12	12
2.	Middle	52	52
3.	Old	36	36
Total		100	100

From the Table 1, it can be concluded that majority of the rice growers were from middle to old age group. Since, we studied rice growers and probably the vounger generation may be occupied elsewhere; the old age group generation is taking care of the farming.

3.1.2 Education

Education helps to broaden the vision of an individual, thereby facilitate in the process of producing desired changes in the behaviour of the people. Information about formal education received by the respondents was collected and presented in Table 2.

The data presented in Table 2 revealed that majority of the respondents (56.00 per cent) had education up to secondary to higher secondary level, followed by 22.00 per cent and 14.00 per cent with up to primary and college level respectively. Very few of them (4.03 per cent) have ability to read and write and only 3.00 per cent were illiterate. It can be concluded that majority of the rice growers had up to secondary level of education and above. This may be due to awareness about importance of education in life.

Table 2: The respondents according to the level of education

N = 100

S. No.	*Education level*	*Number*	*Per cent*
1.	Illiterate	03	03
2.	Can read and write	05	05
3.	Primary education	22	22
4.	Secondary/Higher Secondary education	56	56
5.	College and above education	14	14
	Total	100	100

3.1.3 Use of Information Sources

With a view to studying the sources of information used by the rice growers for getting information related to rice cultivation, the data were collected, analyzed and presented in Table 3. It indicated that major formal sources of information preferred and utilised by the rice growers were fertilizer depot and Service co-operative societies. In case of informal sources like neighbours, friends or progressive farmers were utilised by them. In case of mass media, they exploited TV, News Papers, Radio, Agricultural Exhibition and printed literatures.

Table 3: The respondents according to the use of information sources

N = 100

S. No.	*Sources of information*	*Mean Score*	*Rank*
A	**A Formal Sources**		
1	Fertilizer Depot.	1.81	I
2	Service of co-operative society	1.78	II
3	Village level worker	0.74	III
4	SMS/Sub-divisional officer	0.56	IV
5	Agricultural Research Scientists	0.21	V

contd...

b	**Formal Sources**		
1	Neighbors	1.88	I
2	Friends/Relatives	1.76	II
3	Progressive farmers	0.89	III
c	**Mass Media**		
1	TV	1.67	I
2	News paper	1.41	II
3	Radio	1.32	III
4	Agricultural Exhibition/ fair	1.31	IV
5	Printed literature	0.61	V

3.1.4 Social Participation

An attempt was made to study the degree of association of the respondent with nearby social organizations. The data regarding social participation are presented in Table 4.

Table 4: The respondents according to their level of social participation

N = 100

S.No	*Social Participation*	*Number*	*Per cent*
1.	Low	30	30
2.	Medium	44	44
3.	High	26	26
	Total	100	100

The data revealed that 44.00 per cent of the respondents had medium level of social participation followed by 30.00 per cent with low level and 26.00 per cent with high level of social participants.

3.1.5 Extension Participation

Participation in various extension activities helps an individual to learn more about the technology and also to acquire more information. An attempt was also made to study the level of extension participation of the respondents. The data regarding extension participation are presented in Table 5.

Table 5: The respondents according to level of extension Prticipation.

N = 100

S.No	*Extension Participation*	*Number*	*Per cent*
1.	Low	24	24
2.	Medium	52	52
3.	High	24	24
Total		100	100

The data in Table 5 reflected that 52.00 per cent of the respondents had medium level of extension participation and at the same time an equal per cent (24.00 per cent) of the respondents had low and high level of extension participation. It can be concluded that about fifty per cent of the respondents had medium level of extension participation. This reflects good interest of the respondents to participate in extension activities.

3.1.6 Size of Land Holding

Land holding has been considered as one of the important variables that determines the economic status and potentiality of farmers to go for modern technology. In view of this, farm size of the respondents was studied and data are presented in Table 6. Table reveals that majority (53.00 per cent) of the respondents had big size of land holding, followed by 22.00, 14.00 and 11.00 per cent with medium, small and marginal size of land holding. From this, it can be concluded that majority of the respondents had above 4 hectares of land holding.

Table 6: The respondents according to size of land holding.

N= 124

S.No	*Category*	*Number*	*Per cent*
1	Marginal Size (Up to 1 ha.)	11	11
2	Small Size (1.01 to 2 ha.)	14	14
3	Medium Size (2.02 to 4 ha.)	22	22
4	Big Size (Above 4 ha.)	53	53
Total		100	100

3.1.7 Land under Rice Cultivation

Not only total land holding, but the proportionate area under rice cultivation also affects the information needs of rice growers. Area under rice cultivation in proportion to the total land possessed by the respondents was considered for measuring this variable. The data regarding land under rice cultivation are presented in Table 7.

Table 7: The respondents according to the land under rice cultivation

N = 100

S.No	*Land under rice cultivation*	*Number*	*Per cent*
1	Up to 35.00 per cent	12	12
2	35.01 to 65.00 per cent	24	24
3	Above 65 per cent	64	64
Total		100	100

Data presented in Table 7 revealed that 64.00 per cent of the respondents had above 65 per cent, 24.00 per cent had in between 35.01 to 65.00 per cent and 12.00 per cent of them had up to 35 per cent of their total land holding under rice cultivation.

3.1.8 Irrigated Area to Total Land

It is visualized that a farmer having more area under irrigation may go for more commercialized approach to farming. This may need for more information. The information with regard to irrigated area to total land holding was considered in terms of proportion. The data collected in this regard are presented in Table 8.

Table 8: The respondents according to the proportion of irrigated land

N = 100

S.No	*Proportion of irrigated land*	*Number*	*Per cent*
1	Up to 35.00 per cent	08	08
2	35.01 to 65.00 per cent	26	26
3	Above 65 per cent	66	66
Total		100	100

Data depicted in Table 8 reflect that that 66.00 per cent of the respondents had above 65 per cent, 26.00 per cent had in between 35.01 to 65.00 per cent and 08.00 per cent of them had up to 35 per cent of their total land holding under irrigation.

3.1.9 Annual Income

Annual income refers to the quantum of money earned by farmers during the year from farm and non-farm sources. Data regarding annual income of respondents were collected and presented in Table 9. It can be seen that majority (52.00 per cent) of the rice growers had medium (Rs. 50,001 to Rs. 100,000/-), followed by 34.00 per cent with high (Above Rs. 100,000/-) and 14.00 per cent with low (Up to Rs. 50,000/-) level of annual income. The medium to big size of land holding with majority of respondents under irrigated farming might be the reason for above findings.

Table 9: The respondents according to the level of income

N = 100

S.No	Level of income	Number	Per cent
1	Low (Up to Rs. 50,000/-)	14	14
2	Medium (Rs. 50,001 to Rs. 100,000/-)	52	52
3	High (Above Rs. 100,000/-)	34	34
	Total	100	100

3.1.10 Cosmopoliteness

This psychological character of the respondents was studied to know their tendency to be in contact with sources, outside of their community/area for getting more information. The collected data are presented in Table 10. The data in Table clearly indicate that slightly more than two fifth of the respondents (42.00 per cent) had medium level of cosmopoliteness. This was followed by 37.00 per cent and 21.00 per cent respondents with low and high level of cosmopoliteness, respectively.

Table 10: The respondents according to the level of cosmopoliteness

N = 100

S.No.	*Level of*	*Number*	*Per cent*
1	Low	37	37
2	Medium	42	42
3	High	21	21
Total		100	100

3.1.11 Economic Motivation

It refers to the occupational excellence in terms of profit making and relative value placed on economic ends by the farmers. The data in this regard were collected and are presented in Table 11. The data revealed that 41.00 per cent of the respondents had medium level of economic motivation followed by 34.00 per cent had low level of economic motivation. There were 25.00 per cent of the respondents, who had high level of economic motivation.

Table 11: The respondents according to the level of economic motivation

N = 100

S.No	*Level of economic motivation*	*Number*	*Per cent*
1	Low	34	34
2	Medium	41	41
3	High	25	25
Total		100	100

3.1.12 Market Orientation

The market orientation is such a psychological trait that is associated with his need for market related information. This helps the farmer to analyze the market intelligence to avail better price. The data regarding level of market orientation of the respondents are presented in Table 12. The Table revealed that 45.00 per cent of the respondents had medium level of market orientation. This was followed by 31.00 per cent and 24.00 per cent of the respondents with low and high level of market orientation, respectively.

Table 12: The respondents according to the level of market orientation

N = 100

S.No	*Level of market orientation*	*Number*	*Per cent*
1	Low	31	31
2	Medium	45	45
3	High	24	24
	Total	100	100

3.2 Information Needs of the Rice Growers

An attempt was made to ascertain information needs of the rice growers. This included information needs about variety, land preparation and sowing, fertilizer management, weed management and inter-culturing, water management, plant protection measures, harvesting and post harvest technology, marketing and supportive facts. The data in this regard are presented in Table 13.

Table 13: The respondents according to their over all information needs for rice cultivation

N = 100

S. No.	*Areas of information*	*Mean score*	*Rank*
1	Variety	1.31	V
2	Schedule of water supply by Canal	1.67	II
3	Preparation of Seedlings	1.36	IV
4	Land preparation and sowing	1.22	VII
5	Fertilizer management	1.67	II
6	Weed management	0.71	IX
7	Irrigation management	1.42	III
8	Plant protection measures	1.89	I
9	Harvesting and post harvesting technology	0.88	VIII
10	Marketing	1.89	I
11	Supportive facts	1.21	VII

The data in Table 13 revealed that major area of information needs expressed by the rice growers in descending order of rank were Plant protection measures, Marketing, Schedule of water supply by Canal, Fertilizer management, Water management, Preparation of Seedlings, Variety, Land preparation and sowing, Supportive facts, Harvesting and post harvesting technology and Weed management.

This means that the rice growers gave highest emphasis on market related information, as this information can help them to a great extent to convert their produce in more money. They were also conscious about information on schedule of water to be supplied by canal as well as plant protection measures. The data also reflects that the rice growers have become more cautious about fertilizer management due to new trend of organic rice farming.

3.2.1 Information Needs about Variety

The information needs of rice growers about various aspects of variety were ascertained and recorded in Table 14.

Table 14: The respondents according to information needs about variety

N = 100

S. No.	*Areas of information*	*Mean score*	*Rank*
1	Source of seeds	1.63	I
2	Suitable high yielding variety for the area	1.27	II
3	Rate of seeds	1.12	III
4	Stock of seeds	0.89	IV
5	Characteristics of high yielding variety	0.67	V

The data presented in Table 14 indicate that majority of the farmers have expressed their needs for information about sources of seeds, suitable high yielding variety for the area and rate of seeds. The reason might be that the farmers is convinced to sow good variety but the availability of seed of suitable variety and its' rate are always a dilemma for him.

Table 15: Information Needs about schedule of water to be supplied by Canal

N =100

S. No.	*Areas of information*	*Mean score*	*Rank*
1	Information about water to be supplied in channel before sowing time	1.82	I
2	Advance information about time and date of supply of water in canal for complete crop period	1.27	II

3.2.3 Information Needs about Schedule of Water to be Supplied by Canal

In the area of the study farmers are facing big dilemma of uncertain availability of the irrigation water through canal, thus they are facing problem of crop failure due to lack of expected irrigation water as and when required in required quantity. The Information about water to be supplied in canal before sowing time and advance information about time and date of supply of water in canal for complete crop period were the expected information needs of the rice growers as seen in Table 15.

3.2.4 Information Needs about Preparation of Seedlings of Rice

It is observed that due to lake of technical know-how farmers are not in position to prepare seedlings to transplant rice crop and face problem in getting it as and when required. Some time due to unavailability of needed varieties' seedlings they grow unwanted variety of rice. Thus information need was studied in this regard and data are presented in Table 16. The Table gives us an idea that; How to select site for raising seedlings, Method of preparing bed for nursery, Plant protection in nursery management, Nutrient management in nursery, Irrigation management in nursery and Proper age to select seedlings for transplanting were the major information realized by the rice growers.

Table 16: Information Needs about preparation of seedlings of rice

N =100

S. No.	*Areas of information*	*Mean score*	*Rank*
1	How to select site for raising seedlings	1.31	I
2	Method of preparing bed for nursery	1.28	II
3	Plant protection in nursery management	1.22	III
5	Nutrient management in nursery	1.14	IV
4	Irrigation management in nursery	1.11	V
5	Proper age to select seedlings for transplanting	1.06	VI

3.2.5 Information Needs about Land Preparation and Sowing

The information needs of the rice growers about various aspects of land preparation and sowing were documented and presented in Table 17. It is evident from the data that major areas information needs expressed by the rice growers were land preparation, Soil treatment methods. Place of availability of soil treatment inputs, Seed rate, Price of soil treatment inputs, Sowing time, Depth of sowing, Method of sowing, Spacing, Seed treatment inputs and Gap filling.

Table 17:The respondents according to information needs about land preparation and sowing

N = 100

S. No.	*Areas of information*	*Mean score*	*Rank*
1	Land preparation	1.67	I
2	Soil treatment methods	1.61	II
3	Place of availability of soil treatment inputs	1.56	III
4	Seed rate	1.56	IV
5	Price of soil treatment inputs	1.51	V

contd...

6	Sowing time	1.34	VI
7	Depth of sowing	1.31	VII
8	Method of sowing	1.21	VII
9	Spacing	1.11	IX
10	Seed treatment inputs	0.93	X
11	Gap filling	0.32	XI

The probable reason for high need for soil treatment related information might be due to the fact that the area encompasses high potentiality of irrigation mainly with canal. This resulted in accumulation of salts on the surface of soil, which forces the farmers to learn more about soil reclamation.

3.2.6 Information Needs about Fertilizer Management

Table 18: The respondents according to information needs about fertilizer management

N = 100

S. No.	*Areas of information*	*Mean score*	*Rank*
1	Price of fertilizers	1.82	I
2	Stock of fertilizers	1.78	II
3	Place of availability of fertilizers	1.74	III
4	Name of advantageous chemical fertilizers for rice	1.72	IV
5	Method and time of fertilizer application	1.71	V
6	Nutrient requirements of plant	1.41	VI
7	Calculating the doze of chemical fertilizer	1.40	VII
8	Deficiency symptoms of major plant nutrients	1.32	VIII
9	Bio-fertilizers	0.98	IX
10	Making organic matter from farm waste	0.87	X
11	Organic manures	0.62	xi

The fertilizer management is indispensable for higher yields. Therefore information needs of the rice growers about various aspects of fertilizer management were studied and the results are presented in Table 18. The data reveal that majority of the rice growers shown their interest of information on Price of fertilizers, Stock of fertilizers, Place of availability of fertilizers, Name of highly advantageous chemical fertilizers for rice, Method and time of fertilizer application, Nutrient requirements of plant, Calculating the doze of chemical fertilizer, Deficiency symptoms of major plant nutrients, Bio-fertilizers, Making organic. This means that the respondents know the importance of this input. The information regarding foresaid aspects may help them reducing the total cost of cultivation.

3.2.7 Information Needs about Weed Management

Table 19: The respondents according to information needs about weed management

N = 100

S. No.	*Areas of information*	*Mean score*	*Rank*
1	Chemical weed control	0.93	I
2	Price of weedicides	0.81	II
3	Place of availability of weedicides	0.72	III
4	Trade name of weedicides	0.61	IV
5	Stock of weedicides	0.42	V
6	Hand weeding	0.21	VII

Weed management is important non-monitoring inputs. The information needs of the rice growers about weed management were ascertained and presented in Table 19; the respondents did not express their higher needs for this particular practice of paddy cultivation. All though some of the least important needs expressed were Chemical weed control, Price of weedicides, Place of availability of weedicides, Trade name of weedicides, Stock of weedicides and Hand weeding.

3.2.8 Information Needs about Irrigation Management

Out of the total cultivable area of Khambhat taluka, 60 per cent is under canal irrigation. The rice is major crop in this Taluka of Anand district. Rice is one of the crops which require highest water to survive. As such, the information needs of the rice growers about irrigation management was identified and presented in Table 20.

Table 20: Respondents according to information needs about irrigation management

N = 100

S. No.	*Areas of information*	*Mean score*	*Rank*
1	Schedule for irrigation	1.78	I
2	Critical Stages of Irrigation	1.77	II
3	How to save crop during shortage of water	1.66	III
4	Fertilizer management during irrigation	1.21	IV
5	Method of irrigation	0.41	V

Table 21: Respondents according to information needs about plant protection measures

N = 100

S. No.	*Areas of information*	*Mean score*	*Rank*
1.	Identification, nature of damage and control measures for insects/pests of rice	1.734	I
2.	Identification, nature of damage and control measures for diseases of rice	1.710	II
3.	Price of insecticides and pesticides	1.532	III
4.	Integrated pest management in rice	1.226	IV
5.	Method of preparing solution of insecticides/pesticides	1.056	V
6.	Trade name of insecticides/pesticides	0.903	VI
7.	Place of availability of insecticides and pesticides	0.895	VII

Weed management is important non-monitoring inputs. The information needs of the rice growers about weed management were ascertained and presented in Table 19; the respondents did not express their higher needs for this particular practice of paddy cultivation. All though some of the least important needs expressed were Chemical weed control, Price of weedicides, Place of availability of weedicides, Trade name of weedicides, Stock of weedicides and Hand weeding.

3.2.9 Information Needs about Plant Protection Measures

During rice cultivation many problems of heavy infestations of insects and diseases are being faced by the farmers. Due to this problem many a time, expected production in rice is not possible. There are many incidences recorded of total failure of rice crop in the area of study during past years. The control of pest and diseases is most important and foremost area of information needs expressed by the rice growers as seen in the result of Table 13. Thus detail needs in this matter were studied and presented in table 21. It was observed from the same table that majority of the rice growers have expressed their information needs about identification, nature of damage and control measures for insects/pests as well as diseases of rice crop; price of insecticides/pesticides; integrated pest management and method of preparing solutions of insecticides/pesticides.

The probable reason for information needs about identification, nature of damage and control measures for insects/pests as well as diseases of rice might be that this crop faces major problems in this regard and if plant protection is not done correctly, that may decrease production increase the cost of cultivation.

3.3. Information Needs about Harvesting and Post Harvest Technology

The information needs of rice growers about harvesting and post harvest technology were identified and presented in Table 22. It shows that Proper time of harvest, Ideal thrasher for thrashings rice, How to store rice production, Care after harvesting at farm level and Care during harvesting were major information needs of rice growers about harvesting and post harvest technology.

Table 22: The respondents according to information needs about harvesting and post harvest technology

N = 100

S. No.	*Areas of information*	*Mean score*	*Rank*
1	Proper time of harvest	1.21	I
2	Ideal thrasher for thrashings rice	1.17	II
3	How to store rice production	1.09	III
4	Care after harvesting at farm level	0.92	IV
5	Care during harvesting	0.34	V

3.3.1 Information Needs about Marketing

In India, production is not a problem. Modern Indian farmers have realized that with the adoption of modern practices of crop higher production is possible and as a result of such realization India has achieved significant improvement in food grain production during last few years. The major helplessness, which Indian farmers are facing, is marketing of their products. Until and unless they will not have information regarding marketing intelligence, they will not get reasonable prices of their commodities. Without having good information as regards to where to market, marketing procedure and how to get best price of their produce, they cannot fetch best price for their produce. So far as marketing decision is concerned, the place of market information is inevitable. Information needs of the respondents about marketing were documented and presented in Table 23. The rice growers expressed high information need for almost all the areas of market information. They articulated high need for information on market price followed by quality parameters that affects price and time of market inflow. The fluctuation in price of rice is very common mainly due to time of inflow, quality of rice and its' demand. Further, considerable variations between market yards are also observed. These leads rice growers to confirm high interest in the information regarding market so as to get maximum returns of their produces.

Table 23: The respondents according to information needs about marketing

N = 100

S. No.	*Areas of information*	*Mean score*	*Rank*
1	Market price	1.79	I
2	Quality parameters that affects price	1.72	II
3	Time of market inflow	1.67	III
4	Place of marketing	1.52	IV
5	Marketing procedure	1.41	V
6	Facilities available at market	1.21	VI
7	Value addition	0.51	VII

3.3.3 Information Needs about Supportive Facts

In addition to the key areas discussed earlier, information regarding weather, government policy, credit facilities, crop insurance and subsidies etc., supportive facts were also considered as key areas for information need. Hence, information needs of the respondents were ascertained about supportive facts related to rice cultivation and presented in Table 24. It can be seen that the rice growers expressed their needs for information about subsidies, insurance and government policies related to rice cultivation. The high cost of cultivation might have led the respondents to get information about subsidies. Similarly, high risk associated with the crop may force them to acquire information of insurance. The market price of rice depends greatly on government policies, which lead the farmers to know more about government policies including support price related to rice crop, declared from time to time.

Table 24: The respondents according to information needs about supportive facts

N = 100

S. No.	*Areas of information*	*Mean score*	*Rank*
1.	Weather forecast	0.556	V
2.	Rice related government policies	1.040	III
3.	Credit / loan facilities for rice cultivation	0.895	IV
4.	Insurance of rice crop	1.161	II
5.	Subsidies for rice cultivation	1.339	I

3.3.4 Correlation between Characteristics of the Rice Growers and their Information Need

Personal, social-communicational, economical and psychological characteristics of rice farmers play very important role in exhibiting information need for successful farming. With this in view, efforts were made to study the correlation between personal traits of the rice growers and their information need. To examine this relationship, correlation coefficient (r) was computed, the results of which are presented in Table 25.

3.3.4.1 Age and Information Need

The data presented in Table 25 clearly indicate that information need of the rice growers had non-significant correlation with their age. This might be due to fact that majority of respondents (55.65 per cent) were from the same middle age group.

Hence, the null hypothesis (H_1) in case of age was accepted and it can be concluded that age of rice growers has no relationship with their information needs.

Table 25: Correlation between characteristics of rice growers and information need

N =100

S. No.	*Personal traits*	*Correlation coefficient value*
1	Age	0.011
2	Education	0.141
3	Use of information sources	0.192*
4	Social participation	0.134
5	Extension participation	0.281*
6	Size of land holding	-0.021
7	Land under rice cultivation	0.481*
8	Irrigated area to total land	0.052
9	Annual income	-0.112
10	Cosmopoliteness	0.193*
11	Economic motivation	-0.040
12	Market orientation	-0.114

*Significant at 0.05 per cent level of probability

3.3.4.2 Education and Information Need

The data offered in Table 25 clearly signify that information need of the rice growers had non-significant correlation with their education. This may be due to the fact that irrespective level of education level of the rice growers had information need for rice cultivation remained indifferent. Thus, the null hypothesis (H_0) in case of education was accepted and it can be concluded that age of rice growers has no relationship with their information needs.

3.3.4.3 Use of Information Sources and Information Need

It is evident from Table 25 that information need of the rice growers had significant correlation with their extent of utilization of information sources. This means that those rice growers who utilize more information sources to acquire information regarding rice cultivation, have shown higher need for such information. This may be due to the fact that those who were using various sources of information might have understood importance of information regarding rice cultivation. For this reason, the null hypothesis (H_0) in case of use of information sources was rejected and it can be concluded that use of information sources of rice growers has positive and significant relationship with their information needs.

3.3.4.4 Social Participation and Information Need

The data in Table pointed out that social participation of the rice growers had non-significant correlation with their information need. It means social organizations of villages failed to motivate rice growers to decide difference between what is and what should be for higher production of rice. That's why, the null hypothesis (Ho) in case of social participation was accepted and it can be said that level of social participation of rice growers had no relationship with their information needs.

3.3.4.5 Extension Participation and Information Need

It is obvious from the result in the same table that the information need of the rice growers had significant correlation with their level of extension participation. It means that extension agencies played pivotal role in identifying rice growers' information needs. The probable reason for this might be that extension activities have been considered as an important source for getting information regarding agriculture and certainly by those farmers, who were inquisitive to obtain information, took active part in extension activities. Therefore, the null hypothesis (H_0) in case of extension participation was rejected and it can be concluded that extension participation of rice growers has positive and significant relationship with their information needs of the rice growers about rice cultivation.

3.3.4.6 Size of Land Holding and Information Need

The Table 25 illustrates that information need of the rice growers had non-significant correlation with their size of land holding. It shows that there were all most similar needs of the rice growers with small, medium and big size of land holdings. The probable reason might be that irrespective of the size of land holding, the rice growers always try to get maximum returns from their available resources. Further, the crop being profitable one, the farmers required to be updated with latest information that leads to stability in information need irrespective to the size of holding. Thus, the null hypothesis (H_0) in case of size of land holding was accepted and it was concluded that size of land holding of rice growers has no relationship with their information needs.

3.3.4.7 Land under Rice Cultivation and Information Need

The data accessible in same table suggest that information need of the rice growers had significant correlation with the proportionate land under rice cultivation of their total land. The cultivation of rice crop is uncertain and it requires weighty familiarity about rice cultivation technology including plant protection measures, irrigation management, and fertilizer management. The farmer, who covers big portion of his total land under rice cultivation, transmits more risk for this crop, which leads them to have more information to minimize risk factors. This may be the reason for above finding. As a result, the null hypothesis (H_0) in case of proportionate land under rice cultivation was rejected and it was concluded that rice growers having high share of total land under rice cultivation has higher information needs.

3.3.4.8 Irrigated Area to Total Land and Information Need

It is apparent from the data in Table 25 that information need of the rice growers had non-significant correlation with their proportionate irrigated area to total land. This shows that any increase in irrigated area has no influence over information need of the rice grower. Hence, the null hypothesis (H_0) in case of irrigated area to total land was accepted and it can be concluded that proportionate irrigated area to total land of rice growers has no relationship with their information needs.

3.4.1 Annual Income and Information Need

The data in Table 25 point out that information need of the rice growers had non-significant correlation with their annual income. It means tat there was similar level of interest to have information on rice among rich and poor rice growers. The rice growers, with irrespective of annual income, will always have interest to increase their income that this factor might have led them to have information need for the rice cultivation technologies. Consequently, the null hypothesis (H_0) in case of annual income was acknowledged and it can be concluded that annual income of rice growers has no relationship with their information needs

3.4.2 Cosmopoliteness and Information Need

It is evident from the data in Table 25 that information need of the rice growers had significant correlation with their level of cosmopoliteness. The farmers with high level of cosmopoliteness tend to avail required information, which ultimately leads to less information need. Hence, the null hypothesis (H_0) in case of cosmopoliteness was rejected and it can be concluded that level of cosmopoliteness of rice growers has relationship with their information needs.

3.4.3 Economic Motivation and Information Need

There was information need of the rice growers had non-significant correlation with their level of economic motivation. This may be due to the similarity in term of economic motivation among those entire farmers who were growing rice. Accordingly, the null hypothesis (H_0) in case of economic motivation was time-honored and it was concluded that level of economic motivation of rice growers has no relationship with their information needs

3.4.4 Market Orientation and Information Need

It is noticeable from the Table that information need of the rice growers had non-significant correlation with their level of market orientation. The farmers growing rice crop were all most similar kind of market orientation. It is also obvious that, all the rice growers want information of rice cultivation, irrespective of their orientation to market. For this reason, the null hypothesis (H_0) in case of market orientation was accepted and it can be concluded that level of market orientation of rice growers has no relationship with their information needs.

FINDINGS

1. Characteristics of Rice Growers

1.1 From this research it can be concluded that majority of the rice growers had middle age (52.00 per cent), up to secondary to higher secondary level of education (56.00 per cent),used fertilizer depot, neighbors an television as sources of information, low to medium level of social participation (74.00 per cent), medium level of extension contact (52.00 per cent), big size of land holding (53.00 per cent), 65 per cent of their total land holding under rice cultivation (64.00 per cent), 65 per cent of their total land under irrigation (66.00 per cent) and 50,000 to 10,000 rupees of annual income (52.00 per cent), while slightly more than two fifth of the rice growers had medium level of cosmopoliteness (42.00 per cent), economic motivation (41.00 per cent) and market orientation (45.00 per cent).

2. Information Needs of the Rice Growers

1.1 Plant protection measures, Marketing, Schedule of water supply by Canal, Fertilizer management, Water management, Preparation of Seedlings, Variety, Land preparation and sowing, Supportive facts, Harvesting and post harvesting technology and Weed management.

1.2 Major information needs of rice growers about improved varieties were sources of seeds, suitable high yielding variety for the area and rate of seeds.

1.3 The Information about water to be supplied in channel before sowing time and advance information about time and date of supply of water in canal for complete crop period were the expected information needs of the rice growers relate for the canal irrigation managers.

1.4 How to select site for raising seedlings, Method of preparing bed for nursery, Plant protection in nursery management, Nutrient management in nursery, Irrigation management in nursery and Proper age to select seedlings for transplanting were the major information realized by the rice growers for nursery management in rice cultivation.

1.5 The major areas information needs expressed by the rice growers about sowing of rice were land preparation, Soil treatment methods. Place of availability of soil treatment inputs, Seed rate, Price of soil treatment inputs, Sowing time, Depth of sowing, Method of sowing, Spacing, Seed treatment inputs and Gap filling.

1.6 The major area of information needs related to fertilizer management were Price of fertilizers, Stock of fertilizers, Place of availability of fertilizers, Name of highly advantageous chemical fertilizers for rice, Method and time of fertilizer application, Nutrient requirements of plant, Calculating the doze of chemical fertilizer, Deficiency symptoms of major plant nutrients, Bio-fertilizers and procedure of making organic fertilizers.

1.7 Some of the smallest amounts of information needs expressed by rice growers for weed management were Chemical weed control, Price of weedicides, Place of availability of weedicides, Trade name of weedicides, Stock of weedicides and Hand weeding.

1.8 The rice growers expressed that Schedule for irrigation, Critical Stages of Irrigation, How to save crop during shortage of water, Fertilizer management during irrigation and Method of irrigation were key areas of their information needs regarding irrigation management in rice crop.

1.9 The major rice information needs about plant protection measures expressed by rice growers were identification, nature of damage and control measures for insects/pests as well as diseases of rice crop; price of insecticides/pesticides; integrated pest management and method of preparing solutions of insecticides/pesticides.

2.10 How to store rice production, Care after harvesting at farm level and Care during harvesting were major information needs of rice growers about harvesting and post harvest technology.

2.11 Rice growers articulated high need for information on market price followed by quality parameters that affects price and time of market inflow.

2.12 The rice growers expressed their needs for information on supportive facts like subsidies, insurance in rice crop and government policies related to rice cultivation.

3. Relationship between Characteristics of the Rice Growers and their Information Needs.

3.1 The characteristics like Use of information sources, Extension participation, Land under rice cultivation and Cosmopoliteness were significantly related with their information needs for rice cultivation. In other words rice growers who had better contact with sources of information, extension personals and more area under rice cultivation as well as high level of cosmopoliteness realized more information on various aspects of rice cultivation.

Month *Areas of Information*	*J*	*F*	*M*	*A*	*M*	*J*	*JU*	*AU*	*S*	*O*	*N*	*D*
Land preparation & Soil treatment												
Nursery management for seedling of plants												
Organic manure												
Variety												
Seed, Seed rate, Seed treatment												

contd...

Sowing, Spacing, Gap filling												
Nutrient requirements of plant												
Chemical fertilizer												
Bio-fertilizers												
Fertilizer application												
Integrated pest management												
Insecticides and pesticides												
Control of insects/ pests												
Disease control												
Chemical weed control												
Schedule and method for irrigation												
Time and method of harvesting												
Care during and after harvesting												
Marketing												
Weather forecast												
Rice related Govt. policies												
Credit / loan facilities												
Insurance of Rice crop												
Subsidies for Rice cultivation												

BIBLIOGRAPHY

1. Anonymous (2004) India, International year of rice, 2004, http://www.fao.org/rice2004/en/p6.htm
2. Anonymous (2003) Rice – History, http://www.porkpeople.com/accomrice_history.html
3. Dolli, S. S. and Sundaraswamy, B. (1994). Influence of socio-economic factors on technological gap in cultivation of pulse crop. Maha J.Extn.Edn. 13, PP: 175-170.
4. Hanuman, L.; Mundra, S. N.; Bareth, L. S. (1997). Information need of tribal and non-tribal farmers participated in rural agriculture fair. Maha.J. of Ext.Edn,Vol.XVI, PP:25-29.
5. Meera, S. N.; Jhamtani, A. and Rao, D. (2003). An analysis of agricultural information communication technology projects in India: Implications to the Agricultural Extension System. National workshop on ICT in Agriculture and Rural Development, DA-IICT; Gandhinagar, during 18-19, December 2003.
6. Patel, D.D (2004); Information need of the cotton growers, Unpublished M. Sc. (Agri.) thesis, A.A.U, Anand.
7. Pawar, S. P.; Sawant, P. A. and Nirban, A. J. (2001). Agriculture information needs of neo-literate farmers from Sindhudurg District. Maha. J. Of Ext. Edn. Vol. XX, PP:53-55.
8. Raut, S. S.; Tawde, N. D. and Chaudhari, D .P. (1995). Influence of the characteristics of young farmers on their training needs about cultivation of fruit crops. Maha. J. Extn. Edn. 11 : 292 – 294.
9. Saini, G. S. (1981). Information needs of the small and marginal farmers of Punjab state. Ind.J.Ext.Edn., Vol.XVII(3&4). PP: 101-102.
10. Samantha, R. R. (1977). A study of some agro-economic, socio-psychological and communication variables associated with repayment behaviour of agricultural credit users of Nationalised banks. Unpublished Ph. D. (Agri.) Thesis, IARI, New Delhi.
11. Sankar Rao, A. B. and Reddy, M. S. (1997). Information need of tribal farmers towards mango production technology. MahaJ. of Extn. Edn., Vol. XVI, PP: 78-81.

12. Satpathy, P. K. (1995). Assessment of training needs of potato growers of Kheda district of Gujarat state. Unpublished M. Sc. (Agri.) Thesis, G.A.U., Anand campus, Anand.

13. Singh, S. P.; Hudde, R. S. and Varma, H. K. (1991). Knowledge gap of citrus growers Ind. J.Ext.Edn.32 (1 & 2) PP : 117.

14. Singh, R and Aggarwal, P. L. (1993). Information needs of Punjabi farmwomen. Ind. J. Ext. Edn., Vol .XXIX(1&2), PP:50-54.

15. Stevens, J. B. and Koch, B. H. (1990). Information needs of peach farmers in Silveston Extension Ward. South–African Journal of Agril. Extension. (19), PP: 7-14.

16. Supe, S. V. (1969). Factors related to different degrees of rationality in decision making among farmers. Unpublished Ph. D. (Agri.) Thesis, IARI, New Delhi.

17. Talati, J. P. (1991). Training needs of banana growers of Kheda district of Gujarat state. Unpublished M. Sc. (Agri.) Thesis, G.A.U., Anand.

18. Vaghela, D. M. (1996). Training needs of banana growers in Anand taluka of Kheda district of Gujarat state. Unpublished M. Sc. (Agri.) Thesis, G.A.U., Anand.

3 Extension Management Ability of the Programme Co-ordinators of KVKs of India

ABSTRACT

At present, more than **604 KVKs** working in India have been designed to have vocational training to the practicing farmers, extension personnel and rural youths as well as for all those who want to seek self-employment. The KVK as a grass root level institution has to meet this need, which is so vital for technical literacy and rapid transfer of technology. The Krishi Vigyan Kendra is an organization, which is concerned with social system having several groups of people like; specialists, administrators, technicians, farmers, extension workers, labourers and other beneficiaries. Each group has its own characteristics, attitudes, traditions, values and forms of behaviour. The people for whose benefits, the extension activities are designed may not be readily interested due to multiplicity of factors like; likeness, inhibitions, taboos, suspiciousness, reservations and prejudices. It is, therefore, expected that manager of the KVK should have enough understanding of such social aspects to get best results out of KVKs. The training organizers have to perform multifold roles of planner, organizer, director, communicator, controller, supervisor, coordinator, leader and decision-maker. It was, therefore, felt necessary to examine existing extension management abilities of training organizers of the KVKs. For this, present study on extension management abilities of training organizers of KVK was undertaken with following objectives:

1. To study personal, job related, organizational and socio-psychological characteristics of the training organizers of KVKs.
2. To measure the managerial ability of training organizers of KVKs.
3. To find out the relationship between selected independent variables and managerial ability of the training organizers of KVKs.
4. To estimate the functional relationship between managerial ability and various independent variables.
5. To study the constraints faced by training organizers of KVKs in the management of different activities of KVKs and seek their suggestions.

The present study was carried out in all the eight zones of India, where Krishi Vigyan Kendras are in operation in transfer of technology process. The Ex-post facto research design was applied for the study. Out of total 320 KVKs in India, with five years of functioning experience with well established, full-fledged activities, about fifty per cent (i.e. 160 KVKs) KVKs were selected purposively for the study. A random sample of the total 160 training organizers of KVKs from the different zones of India was selected for this study. The methodological procedure consisted of dependent and independent variables. These variables were measured by using the well developed scales developed by various social scientists with due modifications. The extension management ability of the training organizers of KVK was dependent variable, which was measured by using the scale developed by Patel (2001) with slight modifications. The data were collected by personal interview as well as mail questionnaire. The data thus, collected were classified, tabulated and analyzed in order to make the finding meaningful. The statistical measures, such as percentage, mean score, standard deviation, coefficient of correlation, stepwise regression and standard partial regression coefficient were used.

MAJOR FINDINGS

1. Majority of the training organizers had old age (58.75 per cent), doctorate level of education (71.88 per cent), nuclear type of family (80.00 per cent), up to four members in their family (60.00 per cent), urban background (67.5 per cent), medium level of total experience as a TO (73.76 per cent), medium level of total experience in KVK (51.88 per cent), less exposure of training (76.25 per cent), medium level of span of control (80.62 per cent), medium level of perception of favourable organizational climate (69.38 per cent), medium level of organizational facility (74.38 per cent), medium level interpersonal communication (58.13 per cent), extrovert personality (76.88 per cent), favorable to highly favorable attitudes towards extension work (78.75 per cent), desirable level of workload (60.00 per cent), reasonable favourable perception towards their job involvement (80.00 per cent) and medium level of perception of job stress (73.12 per cent). Whereas, two fifth to half of the training organizers of KVKs had good physical health (40.00 per cent) and medium level of strength of staff provided them (50.00 per cent).
2. Slightly more than half of the training organizers had high level of planning ability (51.25 per cent), majority of them

had medium to high level of organizing capacity (85.00 per cent), high level of ability of directing their subordinates (76.87 per cent), high level of ability of communicating (58.13 per cent), medium to high level of capacity to maintain human relations (91.87 per cent), high level of lead taking behaviour (61.88 per cent), high level of supervising ability (71.26 per cent), high level of coordinating ability (62.50 per cent) and medium level of overall extension management ability (68.12 per cent), whereas slightly less than half of the training organizers had high level of decision making capacity (45.00 per cent) and high level of controlling power (48.12 per cent).

3. The personal variables of training organizers of KVKs such as; young age, higher education, vigorous status of health and rural native place, while organizational variables like; conducive organizational climate, needed organizational facility and better interpersonal communication, whereas socio-psychological variables for instance; extrovert personality and positive attitudes towards extension work and low level of job stress played significant role on their extension management ability.
4. The 45.50 per cent variation in extension management ability of training organizers of KVKs was contributed by attitude toward extension work. However, attitude towards extension work plus personality of the training organizers accounted for 54.20 per cent whereas, attitude towards extension work plus personality of the training organizers plus organizational climate together accounted 55.70 per cent variation in extension management ability of training organizers of KVKs.
5. Lack of promotion channel for the staff of KVK, number of vacant post of the subordinate staff intermittent flow of the funds from organization to KVK head for carrying out mandated activities, unwanted pressure from the local politician to fetch more benefits from KVK schemes to their own jurisdiction, untimely flow of funds for the extension activities and FLDs, frequent transfers of the staff from KVK to other department were the major constraints faced by the

training organizers in managing the KVK. Further they opined that most of the staff has to engage with organizational work or other than KVK activities, lack of adequate infrastructure facilities such as building, fencing and tube well, vehicle, unnecessary involvement of organizational head in taking decision regarding planning, execution and implementations of the KVK activities, inadequate required educational qualification among the subordinate staff, local organizational rules do not allow the KVK staff to work freely according to the guidelines of council were the other constraints experienced and expressed by the training organizers to restrict their extension management ability.

6. The major suggestions given by training organizers in performing their work effectively as extension mangers of KVKs were related to urgent appointment of vacant post, appointment of qualified technical staff, sufficient and timely supply of main and additional grants, provision of sufficient primary and advanced infrastructural facilities, career advancement, restriction on frequent transfer, modern information communication technology, effective linkage with other organizations and personnel management policy.

1. INTRODUCTION

India's population has largely crossed 1 billion and by 1035 experts project that India will overtake China as the world's most populous country. At the same time India already holds first place in another category; the county is home to the world's largest number of poor people. Today estimated 350 million out of 1 billion Indians fall between the poverty line. This number represents a grim reality; while India claims self-sufficiency in production of grains, two-third of the Indian children under the age of 5 are malnourished. Infant mortality is 65 per 1000 live births, compared with 33 for China. Adult literacy rates of 65 per cent for men and 38 per cent for women fall far short of those in China 90 per cent for men and above 73 per cent for women.

Can India achieve its own economic miracle, pushing back poverty and stimulating a vibrant economy with opportunities for its entire people? The task for India's policymakers is to recharge agricultural growth, enabling poor people to contribute their skills and energies to the country's economic activities to get the enhanced benefits of economic success.

Indian agricultural sector is vast and has continued to be the backbone of our economy. It supports 70 per cent of the country's population, which contributes 30-35 per cent of GDP and generates about 20 per cent of export earnings (Mehta, 1998). It provides employment for approximately 62 per cent of the workforce (Singh, 2000). It is indeed heartening that during the 55th year of India's independence, we could reap a record harvest of 208 million tonnes of food grains (Anonymous, 2002).

During the past four decades, spectacular progress has been achieved in agricultural production in the country, which has been well chronicled and universally recognized. In the oil seed sector, there has been a large quantum jump ushering in the yellow revolution increasing production of oilseeds from 11 to 22 million tonnes during the last one decade. The export of oil meal, oil-cake and minor oil has doubled. Likewise the dairy sector has witnessed a white revolution. In inland fisheries, the country has registered an unprecedented compound growth of 10 per cent during the nineties brining blue revolution. The growth in horticulture and poultry production has also been of a higher order (Anonymous, 1999).

In spite of momentous increase in per capita availability of food grain in India, there is no surety of filling the number of empty stomachs multiplied in the years to come. The population of India has already crossed 104 billions and is still increasing alarmingly and that put a great pressure on the food grain production of India. On a rough reckoning it is an acceptable fact that our country has achieved a marvelous success in food grain production from a bare 51 million tonnes in 1951-52 to 212 million tonnes in the year 2003. In spite of this, India is still hungry. It is estimated that India would need to produce an additional 50 million tonnes of food grain by 2010 AD to feed its anticipated population in that year. As per latest information the projected milk production and per capita milk consumption were 80.00 MT and 210 grams per day for the year 2002 (Anonymous, 2001) and more than 100 million tonnes in 2009. Now we should plan for sustaining high milk production as well as increasing productivity of animals. Unless we concentrate our effort on these areas, the shortage of staple foods is likely to be one of the greatest problems. In other words, the need has been arisen for a "Second Green Revolution" in the country through the objective of providing as envisaged in the first "Green Revolution" had been achieved (Paroda, 2000). To achieve the best results from agriculture, Indian farmers will have to make the best use of available modern high production oriented agricultural technologies. This poses a major challenge not only for the policy makers but also more directly to the agricultural scientists, extension workers, agricultural educationist and farmers.

A definite approach to craft India as developed nation in the shortest possible time is realized by a rapid agricultural growth. To accomplish the admired dream of father of nation; Mahatma Gandhi, we have to work with commitment in taking over newer information and technology to our millions of farmers who work hard vigorously in the sun to feed the nation (Anonymous, 2000). The world is shifting very rapidly. Technology of today becomes outdated tomorrow and agriculture is not exception to this. Agriculture plays an imperative role in Indian economy.

The recent advances in agriculture and animal husbandry have demonstrated potential for maximization of agricultural and milk productivity. Today the new developments and techniques have taken over the old ones. However, reality is also that expected numbers of the farmers still do not adopt such recently innovated technologies.

Our former prime minister, Shri Atal Bihari Vajpayee's new slogan "Jai Vigyan" in addition to "Jai Javan" and "Jai Kishan" has provided that the scientific community is not only a sense of pride but it has also big challenge to accomplish the task of doubling the food production during five years to come (Anonymous, 1999)*. Thus, the National Agricultural Technology Project (NATP) with an outlay of Rs. 1000 crores was dedicated recently to the Indian farmers with a firm belief that it would pave the way to doubling the farm production by the year 2010 (Anonymous, 2000)*.

The present age has been rightly termed as an "Information Age". People want adequate and authentic information as early as possible. Farmers as human beings are also anxious and become more desirous with the advancement in science and technology to know what is happening in the field of agriculture. Farmers have enthusiasm to obtain knowledge, particularly in the field of modern agriculture to become psychologically strong and favorable with necessary capacities to adopt modern methods of agriculture. To fulfill the expectations of the farmers, an efficient transfer of technologies is the sole solution.

An extensive extension and training programmes are highly essential to increase the productivity in agriculture. Moreover, many noteworthy changes in the field of agriculture have also been observed recently. These major changes include demand for agricultural product, growing population, economic liberalization, operationalization of GATT agreement, privatization of extension services, sustainability in farming practices, rapid growth in information technology, development of new production technology, role of NGOs and role of indigenous knowledge and experience of farmers. Most of these changes require competent farmers who are able to increase productivity and maintain sustainability of their farming system (Van den Ban and Hawkins, 1996). Therefore, the extension functionaries in coming years under changing scenario will have to play larger role as manager to discharge their responsibility for the fulfillment of the changing requirements of skills and knowledge of farmers (Tripathi, 2000).

1.1 Agricultural Extension System

The productivity level in agriculture mainly depends to a large extent on what type of response a farmer gives to the various plans and strategies of agricultural extension evolved from time to time. The idea behind introducing the various extension programmes is to evolve to good response from the farmers (Singhi *et al.*, 1982).

Agricultural extension is element of the technology transfer system, which is mainly concerned with conveying information and knowledge of important technology from Researchers to Farmers. Broadly, there are four major components of the Indian Agricultural Extension System (IAES) or Transfer of Technology System at present. These components are agricultural extension service, agricultural extension education system, agricultural extension programmes and special rural development programmes.

As far as agricultural extension service is concerned, it is carried out by the state department of agriculture through the training and visit system, as agriculture is a state subject. In agricultural extension education system, the transfer of agricultural information and skill training support to the training and visit system is provided by the Indian Council of Agricultural Research (ICAR) and State Agricultural

Universities (SAUs). The extension system of ICAR and SAU's undertakes first line extension activities, by this system they develop extension methodology, refine and transfer first line technology and provide feedback to scientists. First, the transfer of technology programmes, second, transfer of technology via training by Krishi Vigyan Kendras on the principles of "Teaching by Doing" and "Learning by Doing" to the farming community as well as to the grass root level extension workers and third, training to the subject matter specialists of training and visit system through monthly workshops by ICAR and SAUs are the important approaches of this system. Agricultural extension projects/ programmes are also undertaken by the inputs industries in the public/ private sectors and some NGOs. Special rural development programmes are implemented by the central as well as state governments, which include agricultural extension too.

Today in first line transfer of technology mission, Krishi Vigyan Kendra is a key extension system, which directly deals with the farming communities.

1.2 Krishi Vigyan Kendra

Krishi Vigyan Kendra (KVK) is an admirable project of ICAR for testing and transmitting agricultural technology to link the space between production and productivity and to increase self-employment opportunities among farming communities. Krishi Vigyan Kendra is the grass root level vocational training institute deliberated for bridging the gap between the available technologies at the one end and their application for increased production at the other end. It provides a strong training support for bringing about production break through in agriculture. The main objectives of KVK are to identify the training needs; to conduct production oriented need based training, to organize different non-formal educational programmes and to develop practical training facilities. The specific objectives of KVK however, may be many and varying depending upon the agro-ecosystem of the operational area, needs of the clients, resources available and the thrust required for increasing the economic status of the targeted group. Thus, each KVK is free to develop its specific objectives to gear up the immediate as well as long term needs of the operational area.

It is not an easy task to confirm the innovative character of Krishi Vigyan Kendra. Concerted and continuous efforts have to be made to realize the ideas of Krishi Vigyan Kendra, by the way of imparting tailor made skill training. Therefore, the successful planning and implementation of extension programmes require not only concerted efforts but also a combination of administrative and managerial skills besides a strong political will. The managerial component in the extension programme is quite complex and at the same time crucial too. It is necessary to create a team of 'turned on' managers who have the special sensitively, adaptability and motivation to undertake the task of Krishi Vigyan Kendra's activities. The management knowledge and practice may help in this task (Bhatia *et al.*, 1990).

The frontline extension system of the Indian Council of Agricultural Research (ICAR) presently includes a network of 604 farm science centers commonly known as Krishi Vigyan Kendras (KVKs) and 10 Trainers Training Centers (TTCs). The activities of the KVKs, which were initially confined to farmer's training and On Farm Demonstrations (OFDs), have been further extended to cover in-service training of extension personnel, so as to upgrade their skills in new technologies and the vocational training of farm youths.

The popular demand for establishment of more KVKs in the country is a testimony that the institutional mechanism of the KVK has served society in general and Indian agriculture in particular quite well. Targeted one KVK in each district of India and there will be a provision for more than one KVKs in the same district on demand basis. This is assured by our honorable prime minister Dr. Manmohan Singh in the inaugural speech in National gathering of KVKs at ICAR held during 26th to 28th October - 2005 at ICAR, New Delhi (Anonymous, 2005).

Diversification to horticulture and value addition to agricultural commodities will hold the key for making Indian agriculture have competitive and profitable. Our thrust has to be providing support to disseminate technology in these areas to tap the potential to the fullest extent. Cheaper labour and extensive research and development support are our strengths. The increase in the production of food grains, oil seeds and dairy has firmed up our resolve to exploit the potential in developing horticulture sector (Anonymous, 1999).

For increasing agricultural productivity and ensuring household food and nutritional security, we need to provide new direction to the transfer of technology mechanism. This requires effective linkages among scientists working in different disciplines necessitate institutional restructuring including renewal/reorientation of the existing system to meet the newer challenges.

To meet changing needs, it is essential to create a cadre of "Technology Agent" from the unemployed youths, who are better trained, equipped and committed to serve our farming community, while generating self employment for them. Some KVKs of India have already trained unemployed youths in vocational activities such as sericulture, mushroom production, poultry production, vermicomposting, fruits and vegetable processing, sheep and goat rearing, bio-technology, net and green house management, tissue culture, nursery management etc. It is believed that integration of science with rural development activities would enhance the functional capability and required impact of extension activities. Hence, the integration of KVKs, both for human resource development and frontline demonstration of technologies at the district level would add value to the process of technology transfer on a system made (Anonymous, 1999).

Thus, Krishi Vigyan Kendra is an institutional project of the ICAR to demonstrate technology generation, technology assessment, technology refinements and dissemination in the field of agriculture and allied sectors. The KVKs have been charged now to take up the responsibilities of technology evaluation and impact assessment, demonstration of technology on the farmer's field and update of the knowledge and skill for the farmers.

1.2.1 Genesis of the KVK

The ICAR constituted a committee in 1973 headed by Dr. Mohan Sinha Mehta of Sevamandir, Udaipur (Rajasthan) for working out a details plan for implementing the Krishi Vigyan Kendra. Based on the recommendation of this committee, Tamil Nadu Agricultural University, Coimbatore, started the first Krishi Vigyan Kendra in 1974 at Pondicherry. The success of KVK at many locations created a great demand for establishment of more KVKs in the each district of country. Nowadays, 504 KVKs are established and functioning in the country.

1.2.2 Objectives of the KVK

1. To plan and conduct survey of the operational area to inventorize and characterize physical and human resources with special reference to identify technological graphs and training needs of the farming community.

2. To plan and conduct production oriented need based short and long duration training courses on the campus as well as off-campus with the priority on the weaker and poorer section.
3. To organize farm science clubs in rural areas for young farmers.
4. To develop and maintain the campus farm demonstration unit on scientific lines.
5. To compel all relevant recommendations for the district to the meaningfully utilized in the training courses and the follow up extension programmes.

1.2.3 Achievement of KVKs

At present 504 Krishi Vigyan Kendra (KVKs) are functioning full-fledged in the country in the service of Indian farming. During recent year, these KVKs have given benefits to 6,10,000 farmers and farmwomen, 1,30,000 rural youth and 68,708 in-service personnel by organizing total 30,517 training programmes. To accelerate dissemination process, Kisan Melas, discussion, farm advisory services and diagnostic film shows were arranged. The KVKs also took up On Farm Testing (OFT) of 607 farm technologies. A total of 32,340 Front Line Demonstrations (FLDs), covering 1,19,260 ha of area, were conducted to demonstrate the potential on newly released production technologies on oilseeds, pulses and other crops, besides this, they also carried out 3,337 demonstrations on livestock production and management. At KVKs 7,490.73 tonnes seeds of cereals, oilseeds, pulses and vegetables, spices, medicinal and aromatic plants, ornamental and plantation crops and forest species and 9,58,670 livestock strains were produced and made available to the farmers. Advisory services were also provided to 3,42,634 farmers and other users. Significant impact of KVK technological interventions is also visible in villages in terms of rainwater harvesting, empowerment of rural and tribal women through income generation activities, dairying as sustainable enterprise for women, reclamation of sodic land, off- season vegetable production, backyard poultry and revival of ginger cultivation in different states (Anonymous, 2005)*.

1.2.4 Services of KVK ICT Network

Understanding the importance of information communication technology in dissemination of agricultural information, ICAR has also tried to link KVKs through Internet. Total 250 KVKs have been decorated as e-linked KVKs in first phase all over the country. They are working for the betterment of the country in different ways, such as:

1. Package of practices and recommendation related to locally relevant governmental schemes etc.
2. Online services on localized weather forecasting, market information, latest recommendations on crop, animal, poultry, fishery etc.
3. Interactive voice response system on local weather forecast and market information.
4. Data base on Frequently Asked Questions (FAQs) by the farmers.
5. Sources of availability of inputs such as seed, seedlings, grafts, bio-fertilizers, bio-pesticides and agricultural implements with prices and user instructions.
6. Soil and water testing facilities and on-line reporting.
7. On-line consultancy to farmers through e-mail, voice mail and video-mail.
8. Event calendar of agriculturally important events in the region, training programmes, study tours, exhibitions and other activities.
9. Publication of newsletters and bulletins on KVK website.
10. Soil, water and plant diagnosis services.
11. Other services related to post harvest technology (PHT) and WTO.
12. Self help group and watershed development programme on-line services.
13. On-line marketing.

1.2.5 Management of Krishi Vigyan Kendra

The extension educationists expressed concern over the alarming knowledge practice gap in agriculture in the country. Experts estimated that about 70 per cent of the farm technologies available in India had not yet been adopted by the farmers as well as wide gap existed between potentials of researchers demonstrated on experimental farms and actual outcome at farmer's level (Anonymous, 1998). Thus, an efficient management is crucial for extension organization including Krishi Vigyan Kendra.

For effective management of irrespective nature of the organization, there are some fundamental principles and methods, which guide for effective functioning. The rise of management science and the increasing faith in its ability to solve myriad problems is an outstanding development of modern time.

As viewed by Koontz *et al.* (1980), management is the task of managers to secure and utilize inputs to the enterprise, transforming them through the managerial functions as planning, organizing, staffing, leading and controlling with the due considerations. The situation is altogether different in case of management of Krishi Vigyan Kendra as a non-business enterprise, which is exposed, to serious constraints and handicaps, not encountered in business enterprises.

Krishi Vigyan Kendra is an organization, which is concerned with social system within which, a number of groups of people are working/ involving *i.e.* specialists, administrators, technicians, farmers, extension workers, labourers, other beneficiaries, etc. Each group has its own characteristics *viz.*, attitudes, traditions, values and form of behaviour. The people for whose benefits, the extension activities are designed may not be readily interested due to multiplicity of factors *viz.*, likeness, inhibitions, taboos, suspiciousness, reservations and prejudices (Bhatia *et al.*, 1990)*. It is therefore expected that manager of the KVK should have enough understanding of such social aspects to get best results out of KVKs.

The Revized Mandets of the KVKs:

1. Conducting "on farm testing" for identifying technologies in terms of location specific sustainable land used system.
2. Organize the training to update the extension personnel with emerging advances in agricultural research on regular basis.
3. Organize short and long term vocational training courses in agricultural and allied vocations for the farmers and rural youth's with emphasis on "Learning by Doing" for higher production on farm and self-employment.
4. Organize Front Line Demonstration on various crops to generate production data and feedback information.
5. Working as a Knowledge and Resource Centre for the KVK District.

Mandatory Activities of Krishi Vigyan Kendra:

Based on mandate the following activities are performed by KVKs

1. On farm testing
2. In service training of extension functionaries
3. Vocational training of practicing farmers, farmwomen and rural youths
4. Front line demonstration and other extension activities.

The various said activities are performed every year by the Krishi Vigyan Kendra through six disciplines *viz.*, Extension Education, Agronomy, Horticulture, Plant protection, Animal Science, Agriculture Engineering etc. The specialists can change activities as per location specific need, which are most relevant taking into consideration natural resources and infrastructure facilities of the district.

Moreover, it is a growing realization that an employee today is indeed a new breed. He appears to have different values, needs and motivation than his predecessor. He is a product of knowledge explosion. Thus, managing today's employee is difficult job. Further, the interpersonal relations with group and relationship between the groups also exist in organizations. The groups are mutually interdependent as they perform basically different functions and require different administrative approaches and management practices. A manager who understands the discipline of management will be more effective than a first rate manager with no more than minimum competence in managerial skills and tools.

The training organizer acts as an extension manager of Krishi Vigyan Kendra who is a key person, on whose shoulders, the effective functioning of Krishi Vigyan Kendra depends. The extension manager will have to perform an important task in guiding the change process. Some principles of management science should be considered which may increase the ability and the competence of extension managers as guides to the change process. Thus, many extension management functions have to be done properly by the training organizer of Krishi Vigyan Kendra as the head to get more productivity from the available resources.

1.3 Explanation of Problem

Krishi Vigyan Kendra is leading edged transfer of technology project of the ICAR. It is planned to create able human resource for agriculture development of the country. It helps in creating competent practicing farmers, farmwomen, rural youth and extension personnel. Understanding such importance of KVK, it is necessary to improve the operational quality of the KVK to get desired results. There is a need to analyze quantitative and qualitative functioning aspects of the KVK with a view to getting apprised strength and weakness. The KVK, which is started with set objectives, needs be assessed by evaluating managerial ability of its leader to understand its utility and effectiveness. It is, therefore, necessary to assess the qualities possessed by its heads or training organizers who are the key persons in effective functioning of centers to understand the works done by KVK in terms of agricultural development. Such study helps to locate the strong and weak points of managers of KVKs in overall functioning. It helps to give direction to the continued work besides brining confidence and satisfaction to the workers.

A creative and visionary leader with demonstrated integrity, enthusiasm, motivation, flexibility and confidence to act as training organizer has great chance to show the real impact of the KVK. The training organizer can establish and guide new research directions, lead technical and social research and development proposals, develop new technologies and capabilities and interact with in key clients. The training organizer as a key position holder should try to utilize opportunity to lead centre and offer at the forefront of the emerging technology (Patel *et al.*, 2004).

The effectiveness of the Krishi Vigyan Kendras mainly depends on how it efficiently and successfully managed by its leader or the heads. Generally it is observed that the training organizers of the KVKs are postgraduates with specialization in different subjects of agriculture. They need to have sufficient experience in managing the multidisciplinary faculties of KVKs. The KVK training organizers have to perform multifold roles of planning, organizing, directing, communicating, controlling, supervising, coordinating, leading, coordination and decision-making.

These managerial qualities make them able to perform role as per the mandate of the KVK. By managing all those qualities they have to create conducive condition for effective training. The success of any

KVK is directly or indirectly depends on the various managerial qualities of training organizers. Koontz and O'Donnell (1972) stated that the practice of managing like other practices will be better by making use of underlying organized knowledge and it is this knowledge whether cruder or advanced, whether exact or inexact that to the extent it is well organized clear and pertinent comprises a science. It was therefore, necessary to examine existing status of Training Organizers of the KVKs. With a view to knowing this, a systematic study thought to undertake on "Managerial ability" of the training organizers.

1.4 Scope and Importance of Study

During, The National Conference of KVKs of India, organized at ICAR, New Delhi from 27-28 October, 2005, our honorable Union Minister of Agriculture, Government of India, Mr. Sharad Pawar expressed the salient features and the importance of KVKs in Indian agriculture and country's economy as a whole. He pointed out that after establishing the first KVK during 1974 at Pondicherry, we could establish another 276 KVKs till the end of the IX plan. He assured that GOI is actively considering to sanction atleast one KVK in each district of our country. He believed and expressed that we need motivated workers and vibrant KVKs to double India's food production, oilseeds production to reduce expensive imports of edible oils, harness potential of rainfed agriculture, transformed agricultural marketing and for the grand success of the second phase of the Green Revolution.

These reflect the prime importance of the KVKs in the service of Indian agriculture. The agriculture scenario of the country has changed rapidly by shift in focus from production for home consumption to market oriented production, from staple cereals to cash crops, own labour use to purchased inputs, traditional technology to modern technology, zero marketable surplus to more marketable surplus, labour use to capital use, traditional knowledge based farming to knowledge and skill intensive farming, less risky, low-yielding and less remunerative to high yielding and income and profit yielding farming. In essence, there is a transition going on from family farming to surplus farming for markets and finally commercial farming. These all changes will be only feasible and adoptable through effective and efficient functioning of the KVKs. The National Horticulture Mission will be implemented by KVKs, (Anonymous, 2005)* whether it is public sector or managed by NGOs will be assisted the mission.

Thus, considering major importance, effectiveness of KVKs can be improved by providing valuable functional pattern with sound management upgradation only. To fulfill these dreams, this study will be quiet useful to the concerned persons. It is clearly understood that the managerial effectiveness will lead to organizational effectiveness and further organizational effectiveness will definitely support the national development. KVKs popularly known as "Krishi Mandirs" may be proved as firestone of Indian second "Green Revolution" programme. Accordingly the present study was conducted on the KVKs, which have successfully completed 5 years of functioning. The study was dealt with the management ability of the training organizers of the KVKs in India.

The inferences drawn from this study would be helpful to the ICAR administrators, policy makers as well as zonal coordinating units in taking right decision and a way for improving the existing deficiencies.

The study will be helpful in making necessary changes for increasing the effectiveness of the KVKs in the country. Many factors affect the extension management ability of the training organizer of KVKs, these factors may be of different natures, some may be organizational and social creating an environment where they have to work, while on the other hand, some factors are entirely personal due to their individual psychological make up. The results of the study will be very useful to ascertain the factors associated with management ability of training organizers of KVKs. The information on key role played by some of the independent variables on extension management ability would also be of great use. The constraints faced by training organizers in performing extension management and suggestions offered by them to overcome some of the practical problems will also be useful in structuring effective strategy for future.

1.5 Objectives of Research

The present study was conducted with general objectives of measuring the extension management ability of training organizers as heads of Krishi Vigyan Kendras. The specific objectives of the study were as under:

1. To study personal, job related, organizational and socio-psychological characteristics of the training organizers of KVKs.
2. To measure the managerial ability of training organizers of KVKs.

3. To find out the relationship between selected independent variables and managerial ability of the training organizers of KVKs.
4. To estimate the functional relationship between managerial ability and various independent variables.
5. To study the constraints faced by training organizers of KVKs in the management of different activities of KVKs and seek their suggestions.

1.6 Limitations of the Study

The followings are the limitations of the present study:

1. The study was limited to the Training Organizers of the Krishi Vigyan Kendra only.
2. Only some of the Training Organizers were studied.
3. The study was limited to measure only extension management ability of the training organizers of Krishi Vigyan Kendras.
4. The results in this study were based on the expressed opinions as well as ability to express own personality of training organizers of selected Krishi Vigyan Kendras only.
5. The constraints faced by training organizers may be varied with host institution viz., Government Organizations, NGOs, ICAR's etc.
6. Interpretations and conclusions were not generalized at macro level.

2. REVIEW OF LITERATURE

The main purpose of this chapter is to organize and present findings of the past studies related to the present investigation. It was revealed that very little research work has been carried out about the objectives under study. However, an effort was made to present a brief account of related literature, keeping in view the objectives of the study.

The findings by research workers, experts etc. at different lines in past are presented under following heads.

2.1 Management effectiveness/management ability of the respondents.

2.2 Profile of the respondents.

2.3 Relationship between independent variables and management ability.

2.4 Constraints in management and suggestions for effective management.
2.5 Recent reviewed literature.

2.1 Management Effectiveness / Management Ability of the Respondents

Shah (2006) reported that majority of the internet user teachers were perceived by their students with high level of knowledge about subject matter (72.50 per cent), high level of motivation skill to involve students in teaching and learning process (59.17 per cent), high level of interest in teaching (70.83 per cent), highly punctual (79.17 per cent), highly confident in teaching (69.17 per cent), partially biased nature (90.83 per cent), highly skillful in using audio-visual aids (71.67 per cent) and with medium command on language (63.34 per cent), where as slightly more than half of them were perceived by their students with high level overall teaching ability (55.00 per cent), high degree of ability to control students (52.50 per cent) and with effective speech (50.00 per cent).

Gardebroek and Lansink (2003) studied the estimation of farm productivity differentials using panel data. Results revealed that specialized fig breeding farms that are characterized by high total factor productivity have more buildings and machinery than farms with low total factors productivity i.e. positive relation between farm productivity and managerial ability.

Patel and Patel (2003) reported that majority (70.77 per cent) of the heads of Krishi Vigyan Kendras had medium, followed by 17.69 per cent with low and 11.54 per cent with high level of management ability.

Trip *et al.* (2003) studied the managerial efficiency of commercial green house growers. They said that nowadays the managerial decision making process has been given new attention, both in theoretical studies as well as in empirical research explaining differences in farm results. However, while critical to explaining efficiency managerial ability has been difficult to measure and therefore often ignored. This study attempted to measure managerial ability. It divided the decision making process into four steps: goal formulation, planning, monitoring and evaluation. The results showed positive associations between firm efficiency and the quality of decision making, indicating that this procedure has been potentially successful and it was more towards successful and a move towards successfully measuring a critical input.

Wu-Sx *et al.* (2003) studied the estimation and decomposition of technical efficiency for sugar beat farms and the results indicated that the average efficiency was 0.88 with 45 per cent of the farms in the sample exhibiting full efficiency and efficiency was positively correlated with managerial ability.

Alvarez and Arias (2002) stated that managerial ability has important implications for farm growth and results revealed that increasing farm size while holding managerial ability constant can be an important source of diseconomies of size.

Nathall (2001) reported that the psychology of decision making from a farm management perspective outlines, what psychology others for changing a person's attributes and considers the structure of a research programme aimed at developing methods for improving an individual's managerial ability.

Tolmay and Hernandez (2001) studied the resistance to biotic and abiotic stress in Triticeae and reported that response to the stress situation occurs independently of the managerial ability, skill and resource level of the producer.

Tripathi (2000) observed that the major area for development of management skills for extension personnel is management of man. For the management of man, the skill requirement on one hand is teamwork and on other hand, leadership, coordination, motivation, delegation, clarity of objectives, analysis of resources environment and developing strategic plans, project planning, timely action, review and feedback for individual effectiveness to manage teamwork.

Sah (1999) reported that the extension performance may be disappointing today but it may show resilience with introduction and institutionalization of new and innovative approaches in decision making, financing, provision of extension, building and maintaining linkage with other system and methods to deliver services through hybrid extension model.

Bhandari (1998) stated that the Indian saying "Yatha Raja, Tatha Praja" that is, if a leader is good and effective, every thing will go smoothly, is no exaggerations. But after all that is not easy to bring professionalism of management. At the same time organizational development in true sense is not possible without professionalism of management.

Anonymous (1998)* opined that everybody wants to be the "Best" and "Have Best", but the sources in terms of man, money, machine and material are limited. Thus, arouse the need for management that would results in efficient and effective utilization of various resources to fulfill the objective.

Kareem and Jayaramaiah (1997) studied problems perceived by the state in the planning and implementation of integrated Rural Development Programme as perceived by BDOs and VEOs. He reported that most important problems encountered by them were misuse of assets by beneficiaries, lack of managerial ability among beneficiaries to take care of the assets and for block officials it was lack of adequate transportation to reach beneficiaries efficiently.

Monappa and Saiyaddain (1997) defined an effective manager as one who properly develops in terms of basic intellectual abilities and the predispositions necessary for carrying out the task smoothly.

Verma (1997) reported components relevant for management of extension system viz.; authority structure, delegation of authority, decentralization of authority, control system. Authority with humanity, responsibility and accountability which if applied properly can make authority effective in administration and management.

Bhople and Palaspagar (1996) reported the significant relationship between supervision characteristics and managerial ability as well as Managerial ability and six management functions viz., labour force, live stock numbers and area managed, educational status, length of service and experience Van den Ban and Hawkins (1996) identified the conditions which an extension services must fulfill before it can work well. These are; good communication information, adaptation to changing environment, motivation of staff, flexibility, identification of management problem, participatory leadership, staff development, specialists, generalists, women extension workers and privatization of extension service for accountability. They further indicated that more systematic planning and careful evaluation of objectives / methods/ results were internal parts of effectiveness of many extension organizations. Information produced in evaluation was an important tool for extension organization management.

Shivram and Dalal (1994) studied the effect of managerial ability on farm financial success and reported the relative importance of each management variable to farm financial success.

Unger *et al.* (1995) reported that the adoption of herbicide based conservation tillage in the southern United States Great plains was influenced by managerial ability and economic consideration.

Patel (1993) studied the research management ability of research scientists working at Head of the Department/Station/Project/Scheme in Gujarat Agricultural University. He concluded that the majority (71.67%) of the heads exhibited medium level of research management ability.

Islam and Alli (1991) identified the important sub-areas for training on management of rural development programme officers namely; project planning, implementation, resource utilization, communication with human relation, motivation, reporting, feed back, coordination, cooperation, problem identification, solution, leadership and decision making.

Popat (1991) studied that productivity of subject matter specialist working under T and V system in Gujarat state. He reported that the majority (70.13 per cent) of the SMSs had moderate level of productivity. The field activities got highest mean productivity, followed by training and contact with research.

Prasad (1991) recommended that effective extension strategy should include; use of modern technology tools, training need assessment, new research information, close linkage between all the social scientists, redesigning extension system, use of feed back in planning, serious attitudinal problem, training on learning by doing, group action of NGOs, timely supply of inputs, proper extension methodology and government support for modernizing farming.

Samanta (1991) stated that effective extension education requires adequate human and financial resources with technical and management skills and knowledge associated with new technology to be disseminated to farmers. He also stressed on good personnel management, which is essential for extension programme success.

Bhatia *et al.*, (1990) stated that successful integration of agriculture, industries and service/supportive sector in the integrated rural development programmes depends on the proper planning, organizing, controlling and directing the various activities. The help/cooperation of target groups and implementing agencies are also highly needed. The cooperation of above mentioned sectors and the various agencies is a very difficult and challenging task. Only the management knowledge and practice may help in this task.

Patel (1990) concluded that a good teacher in agricultural college should have a sound and up-to-date knowledge of the subject, possessing at least postgraduate degree in the respective subject and should be methodical in teaching. As far as personal qualities are concerned, he should be honest, punctual in his works, open minded, enthusiastic and having love for his professional as well as the clients.

Chari and Nandapurkar (1989) studied that the managerial ability of the respondents was highly significant with planning, organizing, human relation, supervision, communication, co-ordination and control. The two most important variables were supervision and human relationship, while the two least important were control and organizing.

2.2 Profile of the Respondents

2.2.1 Personal Variables

2.2.1.1 Age

Parmar (2005) revealed that majority (61.67 per cent) of the agricultural scientists were found with middle age group, followed by 38.33 per cent with old age and none of them were with young age.

Sukhadia (2005) found that exactly three-fifth (60.00 per cent) agricultural authors were found with middle age group, followed by old aged (34.55 per cent) and young aged (5.45 per cent), respectively.

Singh *et al.*, (2004) revealed that majority of the respondents were young age group (49.41 per cent), while almost the same percentage (45.88 per cent) belonged to middle age group (36 to 45 years). There were hardly 5.00 per cent having more than 45 years of age. It implies that most of the Horticultural Development Officers were either young or of middle age group.

Patel (2001) indicated that 49.23 per cent of the training organizers of KVKs were found to be in old age group, followed by 34.62 per cent with middle age group while only 16.15 per cent of the Training organizers were from young age group.

Wase (2001) observed the majority of chilli grower managers (52.50 per cent) were in the age group of 36 to 50 years i.e. middle age category.

Borkar and Chimurkar (1996) reported that most of the scientists were in age group of 36 and above.

Patel (1993) indicated that majority (61.67 per cent) of the HODs were in old age group, followed by 33.33 per cent were in middle age group and 5.00 per cent were in young age group.

Patel (1992) observed that majority (60.00 per cent) of the respondents were found to be in the middle age group, followed by young age group (22.00 per cent) and old age group (18.00 per cent).

Popat (1991) reported that majority (74.02 per cent) of the Research Scientist working as Head of Department/Station/Project/ Scheme in GAU were in old age group followed by 22.08 per cent and 3.20 per cent of the respondents were in middle age group and young age group, respectively.

Malik (1988) revealed that majority (56.55 per cent) of the teachers in Gujarat Agricultural University were found to be in the middle age group, followed by young age group (26.26) and old age group (13.79 per cent). He also indicated that more than 80.00 per cent of the teachers belong to young and middle age group.

2.2.1.2 Education

Parmar (2005) found that more than half (57.50 per cent) of the agricultural scientists were educated up to doctoral degree, followed by 39.17 per cent up to post graduate degree, while 2.50 and 0.83 per cent of them were graduates and post-doctorate, respectively.

Sukhadia (2005) indicated that exactly half (50.00 per cent) of the agricultural authors were educated upto M.Sc. degree, followed by 41.82 per cent and 8.18 per cent with Ph.D. and B.Sc. degree, respectively.

Singh *et al.*, (2004) reported that majority (80.00 per cent) of HODs were possessed B. Sc. (Agri). degree, while 20 per cent of them were with M.Sc. (Agri).

Patel (2001) found that majority (71.54 per cent) of the Training organizers possessed doctorate degree, whereas 22.31 per cent and 6.15 per cent Training organizers were found with master and bachelor degree, respectively.

Wase (2001) showed that relatively higher proportion of chilli growers were primary school educated (29.17 per cent), followed by 25.83 per cent with high school and 23.33 per cent with college education, while only 5.84 per cent of them were illiterate.

Borkar and Chimurkar (1996) reported that farm scientists (83.33 per cent) had master degree and only 16.67 per cent possessed the doctorate degrees in their respective subjects of specialization.

Patel (1993) revealed that more than half (55.00 per cent) of the HODs were Ph. D and M. Sc. (36.67 per cent), whereas only 8.33 per cent of them were B. Sc.

Patel (1992) observed that more than half (53.00 per cent) of the respondents had their qualification as doctorate and 47.00 per cent of the respondents were M. Sc. Level.

Popat (1991) revealed that majority (80.52 per cent) of the SMSs working under T and V system in Gujarat State were graduate, followed by 10.19 per cent with agriculture diploma and 9.09 per cent with post graduation.

Khare *et al.*, (1987) reported that the role performance level of SMSs with postgraduate qualification was judged higher than simple graduate SMSs.

2.2.1.3 Health

Patel (1993) indicated that more than half (55.00 per cent) of the research scientists working as Head of Department/Station/Project/Scheme in GAU were having very good health, whereas 38.33 per cent of them were having good health, only 6.67 per cent were having average health and none of them were having poor or very poor health.

2.2.1.4 Size of Family

Patel (2001) found that majority (70.00 per cent) of the Training Organizer of KVK had medium size of family; while equal number of respondents had small (15.38 per cent) and big family size (14.62 per cent).

Patel (1993) indicated that 40.00 per cent of the HODs had medium size of family, followed by 36.67 per cent and 23.33 per cent had small and big size of family, respectively.

Popat (1991) found that majority (62.33 per cent) of the SMSs working under T and V system in Gujarat State had medium size of family, while (32.47 per cent) of them had small size family and 5.20 per cent had big size of family.

2.2.1.5 Type of Family

Patel (1990) stated that majority (64.17 per cent) of the agricultural teachers were belonged to joint family, while 35.83 per cent belonged to nuclear family.

Thakkar (1986) found that majority (76.67 per cent) of the farmers were belonged to joint family.

2.2.1.6 Native Place

Shah (2006) reported that majority of the teachers working in AAU had rural background.

Patel (1993) reported that majority (81.67 per cent) of the Departmental heads were belonging from Gujarat state while remaining (18.33 per cent) of the respondents were belonging from other states.

Popat (1991) observed that majority (77.92 per cent) of the SMSs working under T and V system in Gujarat State were belonged to rural area.

2.2.2 Job Related Variables

2.2.2.1 Experience as a Head of Organization

Patel (2001) observed that majority (62.31 per cent) of the training organizers were in middle level of experience as KVK head followed by low (19.23 per cent) and high (18.46 per cent) level of experience as KVK head.

Patel (1993) revealed that majority (68.33 per cent) of the Departmental head had medium level of service experience as a head, followed by high (16.67 per cent) and low (15.00 per cent), respectively.

Patel (1990) indicated that about one half (49.35 per cent) of the respondents had medium level of service experience, followed by low (27.27 per cent) and high (28.38 per cent), respectively.

Popat (1991) studied the effects of service experience for particular post on productivity. He concluded that experience as SMS has positive but non-significant relationship with SMS's level of productivity.

Veerbhadraiah (1983) revealed that job performance of the Deputy Director was high as compared to Assistant Director of Agriculture.

When performance of 27 specific items was considered individually, the Deputy Directors of Agriculture were observed to perform better than Assistant Directors of Agriculture.

Jaggi (1978) analyzed data for managers with service of five years and less in one category and service over five years in the other category. The results indicated that managers with shorter service would be less authoritarian.

Kholn and Schooler (1973) studied the relationship between man's occupational experience and his psychological functioning and suggested that these results form a continuing interplay between job and man in which the effects of job on man are far from trivial.

2.2.2.2 Total Experience

Shah (2006) reported that Majority (69.00 per cent) of the AAU teachers had high to medium level of teaching experience.

Parmar (2005) stated that 65.00 per cent of the agricultural scientists had medium service experience, whereas 18.33 per cent had less and 16.67 per cent had wide service experience.

Sukhadia (2005) observed that exactly three-fifth (60.00 per cent) of the agricultural authors had medium length of service; whereas 21.82 per cent had low and 18.18 per cent had wide length of service, respectively.

Singh *et al.*, (2004) reported that almost half (49.41 per cent) of the HDOs had reasonable years of experience i.e. 13 to 22 years of service, while 45.88 per cent of them had less than 12 years of total service and remaining 5.00 per cent of them had rich experience.

Patel (2001) reported that majority (60.77 per cent) of the training organizers had medium level of total experience; followed by 23.08 per cent and 16.15 per cent with low and high level of total experience, respectively.

Borkar and Chimurkar (1996) revealed that scientists were having quite a large service experience of more than 16 years, which was spread through all major activities like teaching, research and extension in most of the cases.

Patel (1993) revealed that majority (71.67 per cent) of the department head had medium level of experience, whereas 16.66 per cent and 11.67 per cent of them had low and high level of experience, respectively.

Patel (1992) observed that about three-fifth (61.00 per cent) of the training organizers had medium level of teaching experience, followed by high (21.00 per cent) and low (18.00 per cent), respectively.

Popat (1991) observed that majority (75.32 per cent) of the SMSs had medium level of service experience, while 20.78 per cent and 3.90 per cent of them had low and high level of service experience, respectively.

Chaudhary *et al.*, (1989) found that the service experience and job performance of ADOs were not associated with each other.

Sakaria (1988) reported that length of service of SMSs had positive and non-significant relationship with role performance.

Malik (1988) found that majority (66.21 per cent) of the teachers in Gujarat Agricultural University had medium experience (1.75 to 17.19 years) in research, followed by low experience (17.44 years) and high (16.55 years). The results indicated that two third of the respondents had medium experience in research.

Rao and Mohan (1986) reported that majority (67.23 per cent) of the APAU teachers had average (13.3 to 27.5 years) total experience 16.36 per cent each had high (more than 27.5 years) and low (less than 13.3 years) total experience, respectively.

2.2.2.3 Training Received

Parmar (2005) reported that majority (78.33 per cent) of agricultural scientists had attained national level training, whereas 84.16 per cent of agricultural scientists and 37.50 per cent of agricultural scientists had never attained international and state level training, respectively.

Singh *et al.*, (2004) reported that only 10.59 (per cent) of Horticultural Development Officers did not attend any training in the field of horticulture/vegetable crops, while almost 90 per cent of them had reasonable number of training attended. As high as 83.53 per cent of them had attended 1 to 2 weeks of training and remaining had undergone more than 15 days duration of training.

Patel (2001) observed that majority (76.00 per cent) of training organizers were found in medium level of training received, while 16.20 per cent and 7.60 per cent of them were found in high and low level of training exposure, respectively.

Singh (2001) reported that 32.52 per cent of the women scientists had attended 1 to 5 training courses within the country, while 1.84

per cent had attended 5 to 10 courses. Regarding training in foreign countries, it was found that 16.56 per cent of the women scientists had participated in 1 to 5 training courses abroad.

Patel (1993) revealed that majority (63.33 per cent) of the HODs had received training; out of which 28.33 per cent had attended less training (below average) and 35.00 per cent had attended more training (above average). About 36.67 per cent of the respondents did not receive any type in-service training.

Patel (1992) reported that majority (64.00 per cent) of the respondents were trained, out of which 31.00 per cent had attended less training (below mean), whereas 33.00 per cent had received more training (above mean). About 36.00 per cent of the respondents did not received any type of training.

Popat (1991) indicated that 25.97 per cent of the SMSs did not receive any type of in-service training, while 74.03 per cent of them were trained. He also concluded that, out of trained respondents, 38.96 per cent attended less training (below average), whereas 35.07 per cent had attended more training (above average).

Ardeshana (1989) reported that majority (72.30 per cent) of the teachers had low level of exposure of training, while only 27.70 per cent of the teachers had high level of exposure of training.

2.2.3 Organizational Variables

2.2.3.1 Span of Control

Patel (2001) found that great majority (90.00 per cent) of training organizers had medium level of span of control; followed by 6.15 per cent and 3.85 per cent respondents had high and low level of span of control, respectively.

Patel (1993) reported that majority (86.67 per cent) of HODs had medium level of span of control and 13.33 per cent of them had high level of span of control.

2.2.3.2 Staff Facility

Shah (2006) stated that Majority of the internet user teachers of AAU had average to adequate level of facility of spacious room to sit and work (96.60 per cent), computer (87.50 per cent), over head projector (83.30 per cent), compact disc writer (79.20 per cent), printer (77.50

per cent), internet connectivity (68.30 per cent) and departmental library facilities (60.00 per cent).

Sukhadia (2005) reported that less than half (45.45 per cent) of the agricultural authors had average facilities in their departments, whereas more than one-third (34.55 per cent) had adequate facilities and 20.00 per cent had inadequate facilities in their departments.

Patel (2001) found that 61.00 per cent of the training organizers felt that existing infrastructural facility in their KVKs was medium, followed by low and high level of infrastructural facility perceived by 20.00 per cent and 19.23 per cent, training organizer, respectively.

Patel (1993) found that majority (58.33 per cent) of HODs had perceived medium level of staff facility, whereas 30.00 per cent and 11.67 per cent of them had perceived high and low level of staff facility, respectively.

Patel (1992) indicated that majority (65.00 per cent) of the respondents felt that the feasibilities available to them were reasonable, whereas 21.00 per cent of them felt that the enough facilities were not available properly and 14.00 per cent felt that facilities available in the department were ample.

Ardeshana (1989) reported that majority (73.84 per cent) of the teachers had average facilities available in their departments. Whereas 15.40 per cent had inadequate facilities and only 10.76 per cent per cent had adequate facilities available in their department. He also concluded that there was positive and non-significant association between attitude of the teachers towards their job and facilities available in the department.

Reddy and Jayaramaiah (1990) reported that the facilities and resources were significantly correlated with effectiveness of VEOs.

Soni (1990) reported that majority (56.10 per cent) of the agricultural authors had average facilities in their departments; whereas 23.17 per cent had adequate facilities and 20.73 per cent had inadequate facilities in their departments.

Dakhore and Bhilegaonkar (1988) found that the proper infrastructural facilities had positive influence on the level of job satisfaction of veterinary extension personnel.

Malik (1988) concluded that more than two third (68.97 per cent) of the teachers expressed that the facilities available in their departments were average, whereas 15.86 per cent and 15.17 per cent had possessed adequate and inadequate facilities in their department, respectively.

2.2.3.3 Organizational Climate

Patel (2001) observed that more than half (56.92 per cent) of training organizers had medium level perception of their organizational climate, while an equal number of respondents (21.54 per cent) had low and high level of perception of their organizational climate.

Patel (1993) reported that majority (71.67 per cent) of the HODs had medium level of perception for their organizational climate, while almost equal number of them had low (13.33 per cent) and high (15.00 per cent) level of perception for departmental climate.

2.2.3.4 Interpersonal Communication

Patel (2001) reported that majority (60.77 per cent) of the training organizers had medium level of interpersonal communication, whereas 23.85 per cent and 15.38 per cent of them had high and low level of interpersonal communication.

Patel (1993) found that majority (76.67 per cent) of the HODs had medium level of interpersonal communication, where 13.33 per cent and 10.00 per cent of them had low and high level of interpersonal communication, respectively.

Popat (1991) observed that majority (71.42 per cent) of the SMSs had medium level of interpersonal communication, while equal number (14.29 per cent) of them had high and low level of interpersonal communication, respectively.

Reddy and Sinha (1990) reported that in case of communication between professor and the assistant professor cadre, the communication process was possible through the associate professors but their role was not more than merely handling over the communication material. They are in fact quite seniors and they feel that they should have been an effective agent rather than merely a postal agent.

Samantha (1985) reported that scientists were not fully satisfied with the communication systems on research, personnel and administrative aspects prevailing in the ICAR institutes.

2.2.4 Socio-psychological Variables

2.2.4.1 Personality of the Head

Shah (2006) reported that Majority (74.20 per cent) of the internet user AAU teachers had medium level of perception about their professional development and having good relation among staff members.

Patel (2001) revealed that more than two-third (69.23 per cent) of the training organizers exhibited extrovert personality, whereas 30.77 per cent exhibited introvert type of personality.

Patel (1993) found that majority (86.67 per cent) of the HODs exhibited extrovert personality, whereas 13.33 per cent had introvert personality.

Bhimjiani (1978) concluded that an extension worker; the agricultural extension officers who were free from worries about dealing with the people and who had developed self confidence of working on group situations or as a member of the team were likely to be more successful.

Chaya (1974) concluded that effective teachers had better personality adjustment.

According to Stagner (1956) each person perceives the world in a manner slightly move different than others. These differences are results of unique individual personality and psychological make up. A job, which is satisfying and motivating to one person, may be boring to others.

2.2.4.2 Attitude towards Extension Work

Shah (2006) observed that more than half (58.34 per cent) of the internet user AAU teachers had impartial attitude, 35.00 per cent had favorable and 6.66 per cent of them had unfavorable attitude towards teaching activity.

Parmar (2005) reported that majority (75.83 per cent) of the teachers of GAU had neutral attitude towards publication, followed by unfavourable and favourable with 15.00 and 9.17 per cent, respectively.

Patel (2001) observed that majority (67.69 per cent) of the training organizer had favourable attitude towards extension work, followed by less favourable (16.92 per cent) and highly favourable (15.39 per cent) attitude towards extension work.

Patel (1993) observed that majority (70.00 per cent) of the HODs had medium intensity of favourable attitude towards research work, followed by low (16.67 per cent) and high (13.33 per cent) intensity of favourable attitude towards research work.

Patel (1992) found that majority (65.00 per cent) of the teachers had medium level of attitude towards teaching, followed by low (21.00 per cent) and high (14.00 per cent) level of attitude towards teaching, respectively.

Popat (1991) observed that majority (76.62 per cent) of the SMSs had medium level of attitude towards extension work, followed by high (15.58 per cent) and low (7.80 per cent) level of attitude towards extension work, respectively.

Rani (1985) reported that majority of the teachers were having favourable attitude towards teaching, research and extension. However, she further revealed that only the attitude towards teaching was significantly associated with teaching productivity.

2.2.4.3 Perceived Work Load

Shah (2006) observed that slightly more than half (53.34 per cent) of the internet user AAU teachers had perceived heavy workload.

Sukhadia (2005) indicated that slightly more than half (51.82 per cent) of the agricultural authors had heavy workload, followed by average (46.36 per cent) and only 1.82 per cent had light workload.

Patel (2001) revealed that more than half of the training organizers (58.46 per cent) had perceived medium level of workload, followed by heavy (40.00 per cent) workload and only 1.54 per cent of them had perceived light workload.

Patel (1993) found that majority (71.67 per cent) of the HODs had perception of heavy workload, whereas 26.67 per cent had perception of medium workload and only 1.66 per cent of them had perception of light workload.

Patel (1992) reported that exactly three-fifth (60.00 per cent) of the teachers had average workload, followed by heavy (38.00 per cent) and only 2.00 per cent of them who had light workload.

Ardeshana (1989) indicated that the majority (53.74 percent) of the teachers had heavy workload, followed by 44.62 per cent with average and 1.54 per cent with light workload.

Soni (1990) indicated that majority (54.88 per cent) of the agricultural authors had heavy workload, followed by average (42.68 per cent) and only 2.44 per cent had light workload.

Malik (1988) found that majority (53.10 per cent) of Gujarat Agricultural University teachers had perceived average workload, followed by heavy (37.93 per cent) and too heavy (8.97 per cent). It was found that there was none of the teachers who had very light and light workload. This indicated that majority of the teachers had average and heavy workload.

Rao (1986) reported that majority (57.27 per cent) of the PAU teachers felt that their workload was up to the average degree, where as 31.81 per cent felt that it was heavy, 6.36 per cent said that it was too heavy, 3.36 per cent felt it light and a fraction of them (0.90 per cent) felt that their workload was very light.

2.2.4.4 Job Involvement

Shah (2006) stated that majority (85.00 per cent) of the AAU teachers had medium to high level of involvement in overall internet exposure.

Patel (2001) found that majority (62.31 per cent) of the training organizers were moderately involved in their jobs, followed by high level (20.77 per cent) and low level (16.62 per cent).

Patel (1993) observed that three-fourth (75.00 per cent) of the HODs exhibited medium level of involvement in their job, whereas 15.00 per cent of them had high level of job involvement and only 10.00 per cent had low level of job involvement

Popat (1991) reported that four fifth (79.22 per cent) of the SMSs exhibited medium level of involvement in their job, whereas 14.29 per cent of them had low level of job involvement and only 6.49 per cent had high level of job involvement.

Veerbhadraiah and Jalihal (1983) observed that some common factors like Job centrality, Duty conscious, striving perfection, indifferent response to work, disgusted with work and inconsistent were responsible for underlying job involvement of Deputy Directors and Assistant Directors of Agriculture.

2.2.4.5 Job Stress

Shah (2006) concluded that half (50.84 per cent) of the internet user AAU teachers had medium level of job stress.

Patel (2001) revealed that about two-third of the training organizers (69.23 per cent) expressed medium level of job stress, whereas 17.69 per cent and 13.08 per cent respondents had experienced high and low level of job stress, respectively.

Patel (1993) found that majority (70.00 per cent) of the HODs experienced medium level of job stress, while equal number (15.00 per cent) of them had high and low level of job stress.

Popat (1991) indicated that majority (63.64 per cent) of the SMSs experienced medium level of job tress, followed by low (19.48 per cent) and high (16.88 per cent) level of job stress, respectively.

Igodan (1987) reported that majority (80.00 per cent) of the extension agents of Ohio State (U.S.A.) were found to experience a low to moderate level of burnout, while less than 20.00 per cent of all agents scored in the upper third of the scores indicating a high level of burnout.

Talukdar (1984) revealed that majority (54.90 per cent) of the ADOs experienced medium level of stress in their job lives. A significant number of them (26.00 per cent) also expressed a high level of job stress.

2.3 Relationship between Independent Variables and Management Ability

2.3.1 Personal Variables

2.3.1.1 Age and Management Ability

Shah (2006) stated that the age of AAU teachers was observed negatively significant the efficient involvement of teachers in managing internet.

Patel (2001) reported that age of the training organizer was not significantly associated with their extension management ability.

Shivram and Dalal (1994) observed significant but negative relationship between age and management of adoption process of gram production Technology in Haryana.

Patel (1993) revealed that there was positive and significant relationship between age of the research scientists and their research management ability.

Patel (1992) found that age of the respondents had positive and significant relationship with teaching effectiveness of the teachers.

Popat (1991) observed that age had non-significant relationship with productivity of the subject matter specialists.

Reddy and Jayaramaiah (1990) observed non-significant correlation between age and the productivity of VEOs.

Nikhade and Kitey (1984) found significant relationship between age of VLWS and their performance. Similar findings were also reported by Sharma *et al.*, (1988).

2.3.1.2 Education and Management Ability

Shah (2006) further stated that academic qualification and teaching experience were found non-significant with the internet exposure of the AAU teachers.

Patel (2001) observed that academic qualification of the training organizers was positively and significantly related with their extension management ability.

Jondhale *et al.*, (2000) reported that education of trained and untrained farmers established significant and positive relationship with the managing ability to adopt recommended groundnut cultivation.

Patel (1993) revealed that there was non-significant relationship between education and research management ability of the research scientists.

Popat (1991) reported that academic qualification was not significant with productivity of subject matter specialists.

Padheria (1989) concluded in a study on extension productivity that academic qualification was positively and significantly correlated with extension productivity.

Sharma *et al.*, (1988) revealed that the role performance of Rural Agricultural Extension Officers of T and V system was significantly and positively correlated with their education.

Khare *et al.*, (1987) showed that the academic qualification of SMSs was significantly associated with their role performance level.

Singh *et al.*, (1982) reveled that education of trained farmers of Krishi Vigyan Kendra had positive and significant relationship with their knowledge and management of adoption level of recommended practices.

2.3.1.3 Health and management ability

Patel (1993) revealed that there was non-significant relationship between health of the Research scientists and research management ability.

2.3.1.4 Size of Family and Management Ability

Patel (2001) concluded that there was non-significant relationship between size of family of the training organizers and extension management ability

Patel (1993) observed that there was non-significant relationship between size of family and research management ability of the Research scientist.

Popat (1991) reported that size of family had non-significantly associated with the productivity of subject matter specialists.

2.3.1.5 Type of Family and Management Ability

Sudesh (2005) concluded that there was non-significant relationship between type of family and attitude towards fruit and vegetable training.

Shivram and Dalal (1994) studied the effect of managerial ability on farm financial success and reported that there was positive and significant relationship between managerial ability and type of family of farm growers.

England *et al.*, (1971) reported that in any organization, the attitude of employees towards their work and type of family has a great bearing on the success or failure.

2.3.1.6 Native Place and Management Ability

Shah (2006) reported that native place of the AAU teachers found negative association with their overall internet exposure

Patel (1993) observed that there was non-significant relationship between belonging state of the Research scientist and their research management ability.

Popat (1991) reported that there was no relationship between rural-urban backgrounds of subject matter specialists and their level of productivity.

2.3.2 Job Related Variables

2.3.2.1 Experience as a Training Organizers in KVK and Management Ability

Patel (2001) concluded that there was no significant relationship between experience as a head or the training organizers and extension management ability.

Patel (1993) observed that there was non-significant relationship between experience as a research head of head of the organizations and their research management ability.

Patel (1992) observed that there was significant relationship between teaching effectiveness of teachers and their teaching experience.

Popat (1991) studied that there was no relationship between service experience as subject matter specialist of SMSs and their level of productivity.

Chaudhary *et al.*, (1989) found that the service experience and job performance of ADOs were not associated with each other.

Rani (1985) reported that there was no significant association between experience and productivity of agricultural scientists.

2.3.2.2 Total Experience and Management Ability

Patel (2001) concluded that there was non-significant relationship between total experience as a head of KVK/the training organizers and extension management ability.

Patel (1993) observed that there was non-significant relationship between total service experience of heads of the organizations and their research management ability.

Popat (1991) studied that there was no relationship between service experience of subject matter specialists and their research management ability.

Padheria (1989) reported that total experience had positive and significant relationship with their extension productivity.

Sakaria (1988) reported that length of service of SMSs had positive and non-significant relationship with their role performance.

Khare *et al.*, (1987) showed that the service experience of SMSs was non-significantly correlated with their role performance level.

2.3.2.3 Training Received and Management Ability

Patel (2001) concluded that there was non-significant relationship between training received by the training organizers and their extension management ability.

Patel (1993) observed that there was non-significant relationship between training received by research heads and their research management ability.

Patel (1992) indicated that there was no relationship between teaching effectiveness of teachers and their training received.

Popat (1991) studied that there was no relationship between training received by subject matter specialists and their research management ability.

Padheria (1989) reported that training received by the extension workers had non-significant relationship with their extension productivity.

Sakaria (1988) reported that total training received by the SMSs had non-significant relationship with their role performance.

Rani (1985) reported that training received by the APAU teachers had no relationship with extension productivity.

2.3.3 Organizational Variables

2.3.3.1 Span of Control and Management Ability

Patel (2001) indicated that span of control had non-significant relationship with extension management ability of the training organizers.

Patel (1993) reported that there was no relationship between span of control of the heads of the research organizations and their research management ability.

2.3.3.2 Staff Facilities and Management Ability

Patel (2001) found that there was non-significant relationship between staff facility and extension management ability of the training organizers.

Patel (1993) found that there was non-significant relationship between facility of the staff with head and their research management ability.

Patel (1992) indicated that there was no relationship between facilities available in department of teachers and their level of teaching effectiveness.

Reddy and Jayaramaiah (1990) reported that the facilities and resources were significantly correlated with effectiveness of VEOs.

Ardeshana (1990) concluded that there was positive and non-significant association between the facilities available in the department and the efficiency of the teachers.

Dakhore and Bhilegaonkar (1988) found that the proper infrastructural facilities had positive influence on the level of job satisfaction of veterinary extension personnel.

Rani (1985) observed that facility provided to the APAU teachers had non-significant relationship with the extension productivity.

2.3.3.3 Organizational climate and management ability

Patel (2001) observed that there was non-significant relationship between organizational climate of the training organizers and their extension management ability.

Patel (1993) reported that there was non-significant relationship between organization climate of the heads of the departments and their research management ability.

Padheria (1989) reported that departmental climate experienced by the extension personnel had no relationship with the extension productivity.

Murthy and Somasundaram (1989) reported that organizational climate was significantly related with role performance of VEWs.

Rani (1985) observed that there was non-significant relationship between organizational climate experienced by the APAU teachers and extension productivity.

2.3.3.4 Interpersonal Communication and Management Ability

Patel (2001) reported that interpersonal communication had non-significant relationship with extension management ability of the training organizers.

Patel (1993) found that there was no relationship between interpersonal communication of heads of research organizations and their research management ability.

Padheria *et al.* (1992) reported that infrastructural facility and interpersonal communication were the most important variable affecting scientific productivity of academic staff.

Popat (1991) observed that there was non-significant relationship between interpersonal communication and level of productivity of research scientists.

Reddy and Jayaramaiah (1990) revealed that interpersonal communication was found significantly correlated with productivity of VEOs.

Laharia (1978) observed that the interpersonal communication had positive correlation with the productivity of agricultural scientists.

2.3.4 Socio-psychological Variables

2.3.4.1 Personality of the Head and Management Ability

Patel (2001) revealed that there was negative significant relationship between personality of the training organizers and their extension management ability.

Patel (1993) found that there was non-significant relationship between personality of the heads of research organizations and their research management ability.

Bhimjiani (1978) stated that significant relationship between the personality factor and the performance of the agricultural extension officers.

2.3.4.2 Attitude towards Extension Work and Management Ability

Shah (2006) reported that attitude towards teaching, professional development, job satisfaction, teaching aptitude of teacher and level of aspiration of the AAU teachers were observed non-significantly correlated with their level of internet exposure.

Patel (2001) observed that attitude towards extension work had positive and significant relationship with extension management ability of training organizers.

Patel (1993) observed that there was no relationship between attitude of the HODs towards delegation of authority and their research management ability.

Patel (1992) found that attitude towards teaching had positive and significant relationship with teaching effectiveness of the agricultural teachers.

Popat (1991) reported that attitude towards extension work had no relationship with level of productivity of SMSs.

Malik and Patel (1990) concluded that attitude towards teaching was found significantly related with degree of satisfaction on the job of the teachers of GAU, Anand campus.

Reddy and Jayaramaiah (1990) showed that attitude of VEOs towards T and V system had significant correlation with their productivity.

Murthy and Somasundaram (1989) revealed that attitude of village extension officers towards T and V system was found non-significant with role performance.

Dakhore and Bhilegaonkar (1988) reported that attitude of veterinary extension personnel towards job had positive and significant relationship with degree of satisfaction on their job.

Rani (1985) revealed that attitude of the APAU teachers towards teaching was significantly associated with teachers productivity.

England *et al.* (1971) concluded that in any organization, the attitude of employees towards their work has a great bearing on the success or failure.

2.3.4.3 Perceived Work Load and Management Ability

Patel (2001) revealed that perceived workload by training organizers had non-significant relationship with extension management ability.

Patel (1993) found that workload perceived by the HODs had negative and non-significant relationship with research management ability of the heads of the organization.

Patel (1992) reported that there was no relationship between perception of workload of agricultural teachers and their level of teaching effectiveness.

Padheria (1989) reported that there was no relationship between workload perceived by the extension personnel and the extension productivity.

Rani (1985) observed that perceived workload by APAU teachers had non-significant relationship with their extension productivity.

2.3.4.4 Job Involvement and Management Ability

Patel (2001) found that there was non-significant relationship between job involvement of the training organizers and extension management ability.

Patel (1993) observed that job involvement of the HOD had non-significant relationship with their research management ability.

Popat (1991) reported that there was no relationship between job involvement of the SMSs and their level of productivity.

Murthy and Somasundaram (1989) concluded that job involvement of village Extension officers was found to be positive and significantly correlated with their role performance.

Reddy and Jayaramaiah (1988) reported that job involvement of VEOs was found to be not significant with job effectiveness.

Veerbhadraiah (1983) revealed that there was significant relationship between the job involvement of Deputy Directors and Assistant Directors of Agriculture and their job performance.

2.3.4.5 Job Stress and Management Ability

Patel (2001) revealed that there was non-significant relationship between job stress and extension management ability.

Patel (1993) found that there was no relationship between job stress of research managers and their research management ability.

Popat (1991) indicated that there was no relationship between job stress of subject matter specialists and their level of productivity.

Murthy and Somasundaram (1989) concluded that job stress was found non-significant with role performance of VEOs.

Talukdar (1984) stated that ADOs job stress exhibited no relationship with their productivity.

2.4 Constraints in Management and Suggestions for Effective Management

Patel (2001) reported that major constraints faced by the heads of the farmers training organization were (i) insufficient staff facility especially technical (ii) disciplines and qualification of the technical staff were not maintained (iii) inadequate facility in respect to demonstrations unit/training laboratory as well as infrastructure (iv) insufficient grant and delay in release of grant (v) inadequate power and authority with the training organizers (vi) lack of vehicle facility especially carrying extension work at village level (vii) modern information and communication technology were not fully available with KVKs which resist the modern development in management of extension work (viii) there was communication gap between research institution and extension organizations as well as non-compliance of technical recommendations. Being situated at remote areas, major farmers training centers faced the poor rural infrastructural and educational facilities, which indirectly influenced management of extension work. (ix) work motivation/incentives with human relation angle was found absent in extension field.

They suggested several measures on above constraints viz., (i) vacant posts especially technical including training organizers should be filled up immediately, (ii) discipline as well as qualification should be maintained while appointing technical staff, (iii) demonstration unit/ training laboratory as well as infrastructural premises should be well equipped, (iv) sufficient grant especially for extension activities should be released timely, (v) power and authority should be delegated to the training organizers adequately, considering their responsibilities, (vi) the centers should be restricted to perform only extension tasks, (vii) modern information and communication technology such as computer, internet fax, zerox, mobile phone etc. should be made available, (viii) a strong coordination and linkage on permanent basis should be established between research institutions and centers, (ix) all the necessary infrastructural facility and educational facility for the staff should be made easily available, (x) there should be a provision for work motivation and incentives for the staff.

Reddy and Rao (1998) concluded that scientist commonly expressed constraints about scientific equipment, training and career development, availability of budget in time, library, laboratory, information system, transport and data analysis system in order of priority.

Pal and Singh (1997) observed that extension system is fraught with many problems and misconcepts viz., (i) department of agriculture feels encroachment of its domain by research system, while researches feel gap in transfer of technologies to farmers (ii) lack of operational funds, (iii) ritualistic nature of extension approach (iv) inadequate training for skill upgradation (v) frequent changes in placement (vi) vacant posts in remote areas (vii) lack of incentive structure (viii) low intensity of investment in extension, (ix) reduction in non-salary expenditure and (x) inadequate information, management and evaluation system.

They further suggested several measures on institutional investment and management fronts to improve allotment and cost efficiency of extension systems viz., (i) deviating from past top-down to system oriented bottom-up approach (ii) strong government will to improve management (iii) improve the ability of extension workers (iv) incentive/merit based promotion (v) extension workers accountability (vi) feed back incorporating into performance evaluation (vii) higher plan allocation (viii) to sub-contract private sector for extension services and to attract private investment in extension directly (ix) extension diversification involving private sectors, NGOs, farmer groups, para-extension workers etc. and (x) well structured information and decision support system. Efficiency of these step are contingent upon government will and wherewithal to implement and ability of extension workers to avail them.

Van Den Ban and Hawkins (1996) reported that the constraints which decrease the effectiveness of agricultural extension organizations were; (i) appropriate technology is not available to extent to the farmers (ii) there are no effective linkage between extension organizations and research institutions (iii) field level personnel lack practical training in agricultural technology (iv) extension personnel lack training in extension methods and communication skills (v) extension personnel lack adequate transport facilities to reach farmers effectively (iv) extension personnel lack essential teaching aids, demonstration material and communication equipments and (vii) due to organizational problems, extension personnel have many other tasks besides extension work. They further identified the seven points to be considered by the administrators/managers of extension organizations in order to manage their organizations viz., (i) policy goals through communication and education or combined with other policy instruments (ii) effective communication between extension and research institutions (iii) to manage the planning of extension programme (iv) farmers participation in planning extension programme and in developing and testing of extension message (v) administrator's/manager's stimulation to develop organization through : staff capacity, clear goals, well trained staff, motivation and promotion (vi) good cooperation and effective communication and (vii) available resources.

2.5 Recent Reviewed Literature

Mangala Rai (2006) stated some of the salient features of the progress and achievements in agricultural research, education and extension and the initiatives taken by the council towards enhancing agricultural productivity. During the year, the council was able to established 128 new KVKs raising their number to 504 in the running year. ICAR have been operating partnership and linkages at the national and international level through memoranda of understanding (MOUs/ work plant) for overall development of agrarian country like India.

Pawar (2006) mentioned the council has established a large network of Krishi Vigyan Kendra (KVKs) mainly for technology dissemination, refinement and front line demonstration of technologies. The farmers are benefited from technology demonstrations, training and extension activities undertaken by KVKs. We must address the issues of increasing unemployment, educated unemployment and massive underemployment in agriculture. Unless agriculture becomes economically rewarding, it will be highly difficult to attract or retain youth in farming.

Chitra Pathak *et al.* (2005) concluded that agriculture extension in the current scenario of a rapidly changing world, has been recognized as an essential mechanism for delivering knowledge (information) and advice as an input for modern farming. Besides performing the conventional task of transferring technology packages to the farming community. Hence, this era of new technological development training of both the extension functionaries and the farmers in the use and application of appropriate technologies should receive topmost priority. She further stated that the large network of front line ICAR – SAU extension system consists of 415 KVKs, 10 trainer's training centers (TTCs), 70 institution village linkage project centers and 44 agricultural technology information centers. Through this network training of 6.0 lakh farmers and farmwomen every year in agriculture and allied fields such as crop production, plant protection, livestock production and farm machinery and tools, and science is competed in the country, with limited resources at the disposal of SAUs and KVK in the sates, the effectiveness and productivity of training programmes are crucial for achieving the desired results.

Kalra and Mishra (2005) reported that we have still a very long way to go and a very active trial and error process has to be run a full course in order to evolve the desired institutional arrangements at grass root level. Scientists have to shoulder the main responsibly regarding arranging funds, planning and implementation of the project. Project could be successful in Krishi Vigyan Kendra (KVK) situated in the village and villagers take up the joint responsibility. Krishi Vigyan Kendra (KVKs) are the best institutes for effective transfer of technologies.

Singh and Padaria (2004) reported that superimposed on TOT approach was the T and V system with emphasis on the role of extension in technology transfer to encourage utilization or research results. The T and V approach was somewhat like a campaign approach an succeeded to some extent where the clients need and technology potential matched. The system was found to be too narrow in its approach and not suitable for small farms and rain-fed areas. In between, the ICAR launched its programmes namely operation research project (ORP), Krishi Vigyan Kendra (KVK), and Lab to Land Programme (LLP) with the same objective in mind. All these programmes were later merged under the programmes of KVK. These programmes did receive some success but not upto the mark, particularly ensuring people's participation. The farming system research and extension programme was also started through KVKs. We have a strong research base of ICAR and university system and a large network of extension system. KVKs are the best institutes at grass root level for effective agricultural extension.

Lee and Kim (2004) showed that the E-linked KVKs can serve the problems for young and well educated people, working in agriculture it is easier to find employment elsewhere than for older people. Hence, a country with a rapid economic growth during last 40 years has 51 % of the farm operators were over 60 years and only 60 % were less than 40 years.

Swanson (2004) reported that the KVKs can train the extension personals for updating their knowledge with changing scenario of the agriculture at global level. There are large differences between countries in the opportunities farm families have to send their children to a vocational agriculture school. In India, a small proportion of these youngsters can get a course of one week of a farm science centre (KVK).

Sulaiman and Hall (2004) stated that it is indeed important that farmers are well informed about new production technologies, which may be profitable in their situation. These technologies are not only developed in government research institution but increasingly also in commercial companies selling inputs to farmer or marketing their products and by innovative farmers. Extension services should use all available information sources about new technologies. They further stated that the development of useful new production technologies requires partnerships between research institutes, companies selling inputs or marketing products and farmers. Unfortunately in many developing countries these partnerships do not exist or are weak. In a few Indian, districts these partnerships are promoted in a world bank project on agricultural technology management agency (ATMA). Krishi Vigyan Kendra has a prime importance in ATMA project.

Van den Ban (2004) concluded that in India thousands of agricultural graduates are trained for IT related job. They can get subsidy from the government for the necessary investments in ICT and make a living from selling information for which farmers feel a need. Agri-informatics and Bio-informatics at KVK is the advance step of this line.

Van den Ban (2004) showed that insufficient vocational agricultural education makes it difficult for Indian farmers to compete in the world market with farmers from other countries. It shows the importance of the farm science centres (KVKs) in Indian agriculture.

Vinayangam and Samanta (2004) stated that the organisations responsibility is to provide an congenial climate and satisfying environment in which people can gain overall satisfaction in their quality of work life (QWL) and help organization to attain its objectives.

It was also stated that if employee has a positive perception of QWL in the organization, he or she will strive to further improve the working conditions and increasing productivity. This is more so, and very much applicable to an organization like 'Farm Science Centre" popularly known as 'Krishi Vigyan Kendra (KVK) established by Indian council of Agricultural Research (ICAR). The work environment of trainers in KVK is pivotal to make themselves fit into the system which generates motivation, commitment and dedication to transfer the technological information and skill to the farmers through training. The study was under taken to find out the perception of trainer's of KVK, working under four organizations namely ICAR, SAUs, State Department of Agriculture and NGOs.

Das (2003) reported that the Krishi Vigyan Kendra apart from having the internet access from the Pravara network also has the Agricultural Research Information System (ARIS) call set-up under the NAPT project of ICAR for access of information from the ICAR research centers and SAUs. The Krishi Vigyan Kendra at Babhaleshwar has internet network linking all the subject matter specialists (SMS) with all the villages in the network and to the internet as well. All the SMSs of the KVK frequently build up the information repository for agriculture extension at its server, which is available for all those connected to the wireless LAN in the Pravara Network. He also stated that a programme to establish a strong e-linkage with the KVKs of 200 districts and technology generating institutions has been prepared for availability of technology information. The objectives of the project are to establish a strong e-linkage with the KVKs and technology generating institutions like ICAR institutes and State Agricultural Universities for availability of technology information for the farmer at KVK level, to enable the KVKs for global access and to act as gateway for availability of such information at local level and to develop necessary human resource for handling cyber extension. The KVK will be equipped to enable them to convert the existing knowledge base into web-enabled electronic form to serve as information hub of the area.

3. THEORETICAL ORIENTATION

The main objective of this chapter is to present schematically the conceptual frame of reference used for this study. An attempt has been made to portray systematic framework, which would provide guidance in selection of variables. The chapter is presented in the following sections:

3.1 Extension management ability as a concept
3.2 Theoretical perspective
3.3 Independent variables
3.4 Derivation of hypotheses
3.5 The conceptual model

3.1 Extension Management Ability as a Concept

Management and administration have been used consciously as synonymous concepts in management literature. Management has been allowed almost as synonymous of administration.

The term 'Management" stems from the word "Manage" which is derived from French word "Manage" meaning 'House keeping'. In Latin literature, the term 'Management' is viewed to be derived from the word 'Manus" which means 'Hand" and it originally was referred almost exclusively to the handling and training of horses.

In general usage, the 'Management" identifies a special group of people whose job is to direct the efforts and activities of other people towards common objectives.

Irrespective of the nature of organization, there are some fundamental principles and methods which guide their effective functioning which in modern terminology we call it as 'management'.

According to Drucker and Fayol (1954), there is no much distinction between the terms 'management' and 'administration' and both are synonymous and used interchangeably. According to them, running a business or social institution or government office, person heading them requires specialized skill. In business, this specialized skill is called 'Management', while in government and social institution this specialized skill is called 'Administration'. But, a question is often asked whether the task of administration is the same as that of management. To probe further into the matter, we thus need to turn to common practices in the government institutions and industries. The distinction between the scope and connotation of the two terms administration and management is not qualitative. It lies in the emphasis. In case of administration, the emphasis, which tends to be wider, long ranged, diffused, comprehensive and more fundamental. While in case of management, the emphasis is more time bound, concreted, delimited, economically viable and more susceptible. The beneficiaries of the administration are at large in entire society, which are neither identified nor homogenous, while the beneficiaries of the management are a few *i.e.,* the owners, the employees and indirectly the consumers and almost homogenous. Generally in administration, long term social costs and social gains occupy a central position, while in management the profit or monetary or economic return on efforts is almost necessarily in the forefront.

According to one school of thought, administration and management are different functions. This view is also supported by eminent writers like Oliver Sheldon, William R. Spriegel and G. W. Milward. The differences between them are administration is concerned with the functions as policy making, legislative, determinative and thinking relates to the apex and top level management and needs administrative ability rather than technical ability. While, management is concerned with the functions as policy implementation, executive governing and doing, relates to the middle and lower level management are required technical ability more than administrative ability. According to E. F. L. Brech (1972) management is a comprehensive generic term, which includes administration as a branch of it. The management function has been divided into two categories. The upper level management is usually called as administration management and the lower level of management is known as operative management. According to V. L. Srivastava (1981), management is a social process entailing the execution of over all polity and plan i.e. functions of execution being also called proper management. While administration is that part of business management which is entrusted with the task of the determination of policy i.e. functions of direction being also called top management.

From the above views, it can be said that the management is a specialized set of functions/principles, which also includes the administration aspects. As the present study is limited to management aspects and viewed on management as a concept expressed by various thinkers are presented below.

According to Davis (1951), management is the function of executive leadership anywhere.

Peter Drucker (1954) defined management as a multipurpose organ that manages a business and manages manager's workers and works.

To, Niles (1956) good management or scientific management achieves social objectives with the best use of human and material, use of time and with satisfaction to the participants.

Vance (1956) defined management as simply the process of decision making and control over the actions of human beings for the purpose of attaining pre-determined goals.

Mee (1963) defined the term management as art of securing maximum results with minimum of efforts so as to secure maximum prosperity and happiness for both employer and employee and thereby giving best possible services to the public.

Terry (1968) defined management as a distinct process consisting of planning, organizing, actuating and controlling personnel to determine and accomplish the objectives by the use of people and resources.

Appley (1980) defined management as guiding human and physical resources into a dynamic and hard hitting organizational unit that attains its objectives to the satisfaction of those served and with a high degree of morale and sense of attainment on the part of those rendering the service.

According to Dasgupta (1969), management is the creation and control of technological and human environment or an organization in which human skills and capacities of individuals and groups find full scope for their effective use in order to accomplish the objectives for which an enterprise has been set up.

Morell (1969) viewed that management is that activity in an organization, which consists in deciding upon the means by which the goals are to be effectively reached.

Hodge and Johnson (1970) defined that management is the process of making decisions and issuing commands on behalf of an organization taking into consideration the complex of objectives, laminations and standards underlying the production and distributing of value required to satisfy members' needs.

Koontz and O'Donnell (1972) defined management as the art of getting things done through and with people in formally organized group, the art of creating an environment in such an organized group where people can perform as individuals and yet cooperate towards attainment of group goals and the art of optimizing efficiency in effectively reaching goals.

Management, according to McGreger (1971), is responsible for organizing the elements of productive enterprise i.e. money, materials, equipments, people etc. in the interest of economic ends.

According to Scanlan (1973), management involves the coordination and integration of all resources both human and technical to accomplish specific results. The five basic functions which have historically formed the core of studying management are planning, decision making organizing, directing and controlling.

Managerial process, according to Hellriegel and Slocum Jr. (1975) is concerned primarily with the design, planning and control of the organization's behaviour as well as the decision making which is required.

As view by Koontz *et al.* (1980), management is the task of managers to secure and utilize inputs to the enterprise, transforming them through the managerial functions of planning, organizing, staffing, leading and controlling with due considerations of external variables to produce outputs.

In these ways, many scientists have contributed to the meaning and concept of management in their own ways. Finally one can be define management as "the effective and efficient integration and coordination of resources to achieve desired objectives." The term extension management can be defined as an effective and efficient utilization and coordination of available resources to achieve the objectives of extension including Krishi Vigyan Kendras.

The term 'Ability' usually denotes a potential for performing some task. A person's ability to perform a task refers to the degree to which he possesses all the psychological attributes necessary for a high level of performance. Performance of an individual is the function of certain characteristics of the individual including his knowledge, skills, motivation, attribute and certain aspects of environmental situation (McGreger, 1967). This can be extended to training organizers of Krishi Vigyan Kendra also. Extension management ability of training organizer is thus, a function of his personal attributes, socio-psychological characteristics and the work environment.

To meet the needs of all the potential beneficiaries, training organizer has to develop better coordination of various extension systems and also has to share the responsibility of extension education with other agencies and organizations. All this calls for new policy initiatives, Creative management of the extension systems and organized efforts to upgrade the knowledge and skills of extension functionaries. Thus, a training organizer has to perform a variety of tasks related to extension management.

3.2 Theoretical Perspectives

The present belief of management or management theory or management science in the current situation can be understood in a better way if we have some idea about its evolution. This section is devoted to a critical review of some comprehensive and refractory hypotheses of management or management thought which is classified and presented into three phase's viz., Pre-scientific Management Approaches, Early Management Approaches and Modern Management Approaches.

3.2.1 Pre-scientific Management Approaches

3.2.1.1 Ancient Management Thoughts

Ancient Greek literature points to the existence of administrative apparatus like council, courts and boards of military personnel to ménage the affairs of the state.

Socrates, a famous philosopher was among the first to look upon management as something different from mere technical knowledge and skill.

The success of the Roman Empire depends on the ability of the rulers to organize their forces according to the scalar principle, neatly defining the area of authority and responsibility of each official.

3.2.1.2 Management of Roman Catholic Church

A set of well defined objective, an effective and efficient organization setup to achieve objectives, division of work based on specialization, employment of staff officers to aid and advice the line managers were some of the striking features of the management of Roman Catholic Church.

3.2.1.3 Managerial Practices by Military Organizations

Military organizations also contributed in their own simplistic way to the development of managerial practices though there was little area of theory in it. Even, so, their techniques of authority relationship between individuals and groups of individuals, direction, motivation and communication underwent considerable improvement over the years. Of late, military organizations have also experimented quite successfully, with the staff principle, according to which persons with specialized knowledge were appointed to aid and advise the field officers.

3.2.1.4 Management Theory by Cameralists

Contribution of management theory was also made by a group of German and Austrian administrators and thinkers, collectively called as the cameralists, who believed in organization of material wealth to make the state strong. The cameralists felt that to achieve this objective, the state needed to adopt the same techniques as an individual. Among the management principles formulated by the cameralists were; (a) job satisfaction (b) proper selection of personnel (c) division of work based on specialization and (d) effective control.

The above concepts were mostly related to the principles of specialization, selection of subordinates and their training and simplification of administration procedures. However, the various ideas stated by them have created awareness about managerial problems.

3.2.2 Early management Approaches

3.2.2.1 Management by Custom School

Exponents of this school of management though emphasize the importance of a study of the experience of successful managers or a study of mistakes made in management. They claimed that such study would provide a better understanding of the most effective way of managing an enterprise.

3.2.2.2 The Decision Theory

Exponents of this school of though emphasize the decision making aspects of management.

3.2.2.3 Scientific Management Movement

Taylor (1915), due to his most significant contribution to the concept of scientific management was regarded as "The father of scientific management movement". He stated that scientific management in its essence consists of combination of great principles of management viz., the development of true science, scientific selection of workers, intimate and friendly cooperation between the management and their workmen. It was based on a systematic study and analysis of various aspects of work involved with a view to finding out the best way of doing a thing, we call it scientific management.

Gilbreth (1915) is regarded as "The father of motion study" and laid the entire foundation of our modern application of job simplification incentive wages plans, work standards and better work methods.

Gilbreth (1915)* emphasized the human factor and stressed physiological effects of fatigue.

Gantt (1919), contribution to scientific management includes task and bonus approach for remunerating workers, indicating a more humanitarian approach and emphasis on service rather than on profit.

Thus, scientific management emphasized faster and better methods of production and also increased out put and efficiency.

3.2.2.4 The Management Process School

The operational or management process fathered by Henry Fayol (1949) emphasized that management is an operational process and can best be studied through an analysis of the managerial functions. The process can basically be divided into five functions namely; planning, organizing, staffing, motivating and controlling although different authors use different words to describe these functions or divisions.

Koontz and O'Donnell (1972) divided the process into planning, organizing, staffing, directing and controlling. Terry (1968), preferred the word actuating to motivating. Gullick's (1925), formula of functions or process of management involves planning, organizing, staffing, directing, coordinating, reporting and budgeting (POSDCORB): Urwick and Gullick (1925) synthesized the ideas of various writers on the process management theory and indicated the probability of the science of management. Many principles of process theory of management have been applied with great success even today in all types of settings, though with some modifications.

3.2.2.5 Human Relation Movement

Elton Mayo (1932) "The founder of human relations movement", stated that the employee's morale had a great influence on productivity and them anager should treat them as social beings instead of economic beings or simply as cogs of wheel. The manager should understand group. The contributions of human relations remain and are being applied even today by managers.

3.2.3 Modern Management Approaches

3.2.3.1 Neo-classical Theory

Neo-classical theory consists of two stream viz., human relations and behavioral science approach. The important elements of this theory are; individual work groups, participative management, orientation, motivation, leadership and employee development. In this approach human factor and values are given more importance.

3.2.3.2 Social System Theory

Barnard Chester (1938) stated that the management is a social system and a system of cultural relationships. Cooperation and group spirit among members are necessary for achievement of organizational objectives. It is very much valuable to the practicing managers.

3.2.3.3 Quantitative Approach

Quantitative approach to management is also known as operational research in management science. This approach offered a systematic analysis and solution to many complex problems faced by management. For decision making on complex issues, mathematical and statistical tools are applied.

3.2.3.4 System Approach

Management is a system, which is a set of interconnected and interrelated elements/components/parts to achieve certain objectives. It has a number of sub-systems and classified into two groups viz. closed system and open system.

3.2.3.5 Situational Approach

According to this approach, different situations demand different solutions. There cannot be universal principles of management appropriate to all situations. This contingency or situational approach helps the management in finding solutions to the problems in specific situations.

3.2.3.6 Modern Management Thoughts

In the recent past, a number of persons have contributed to the development of management science. More important among them are Oliver Sheldon, Mooney and Railey, Chester Barnard, Lyndall, Uriwick and Peter Drucker. Peter Drucker outshines all other contemporary thinkers and he is considered as "The father of Modern Management Thoughts". Drucker considers management as a mere practice rather than a science. He advocates centralized control and decentralized structure.

Peter Drucker (1974) stated that every employee has crystallized and complex sets of needs, values, way of perceiving the world and his personality, which he brings with him to his work place. He is thus, not a raw mould. He tries to fit his psychological make up with the climate of the organization. The fact that his psychological characteristics influence his behaviour, suggests that his job behaviour and his motivation for job performance will also be influenced by these characteristics. That means, in hiring a worker, one always hires the whole man and this explains why the improvement of human effectiveness in work is the greatest opportunity for improvement of performance and output of the organization.

Vroom (1964) also emphasized that job performance is a result of interaction between worker's skill, ability and motivation.

3.2.4 Implication of the Theories

On the basis of the different theories discussed above, it could be inferred that the various functions/principles or practices of management for any organizations alongwith certain socio-economic forces operative at a given time provide a frame for individual to work effectively. The processes or practices of extension management and socio-economic forces may either enhance or restrict or at least lower the pace of the individual, the training organizer in the present study. The socio-economic dynamics enables him to be more able in extension management or restricting him to less extension management ability level.

These theories further lead to identify certain elements of management ability. These factors could be analyzed at various levels viz., management functions/skills/practices, administrative management aspects, psychological aspects, motivational aspects, social system aspects, human relation aspects and many other related aspects viz., planning and organizing the activities, proper direction of work, effective communication, control over work, human related aspects of staff, leading the entire programme of work, providing proper supervision to the subordinates, coordinating the programme activities in a well manner and take proper decision at various levels, were emerged as important indicators for extension management ability of training organizer of Krishi Vigyan Kendra and were felt appropriate for inclusion to understand extension management ability of training organizers.

The indicators as well as sub indicators of training organizer's extension management ability derived on the basis of different theories were subjected to judge's opinions after through considerations of management experts.

3.3 Independent Variables

The role of situation or environment is very crucial in understanding human action. Behaviour takes place in a situation and it has therefore, profound influence on the individual's action. Sometimes, situations "present" the actor with certain goal to pursue; Rogers (1962) therefore confirmed the role of situation on determining an action of individuals.

It therefore, implies that the social system, in which individuals are members, has dominant effect on the behaviour of individuals. Members of social system characterized by traditional norms can be proven to resist change. This is a much widely accepted generalization.

Human action is dependent upon many factors. As conceptualized by Pearsons (1954), the action takes place in a situation consisting of social economic, physical and cultural factors and socio-economic dynamics operative at a given time. What action, an actor performed depends on how he orients to and evaluates the given situations i.e. the alternatives available and the expected gratification to be derived. The orientation and evaluation processes in turn depend on the personality and cultural systems. To be precise the action process is dependent upon personal, economic, social and psychological factors possessed by individual. Hence the analysis of action involves the concept and theories of various disciplines. He also mentioned that the selection of goals as well as means depends on such factors as cultural norms and values i.e. social, biological, capacities and accessibility to the physical and social environment.

On the basis of the previous discussion certain causative variables *viz.*, age, academic qualification, health, type of family, family size, native place, total working experience in KVK, experience as KVK head, total training received, availability of technical staff, span of control, interpersonal communication, organization climatic, organizational facility, infrastructural facility, extension staff facility, personality, perceived workload, attitude towards extension work, job involvement and job stress were selected to know their influence on extension management ability of training organizers of Krishi Vigyan Kendras.

3.3.1 Operationalization and Definition of the Independent Variables

3.3.1.1 Personal Variable

1. **Age:** It refers to number of chronological years completed by the respondent on the date of interview.
2. **Education:** It is formal education level of respondents.
3. **Health:** It is the physical and mental well-being of TOs as perceived by them in terms of very good health, Good health, Average health, Poor health and Very poor health, it was measured as method used by Popat, (1991).
4. **Type of family:** It refers to type of family of TOs in terms of nuclear or joint type of family.
5. **Family size:** It refers to number of family members in the family of training Organizer.
6. **Native Place:** In refers to the birth place of the training organizer. It is rural or urban background, surroundings, conditions and environment of the training organizers where he grew up.

3.3.1.2 Job Related Variables

1. **Experience as a training organizer:** It is number of years of experience of the respondents as training organizers or in-charge of any KVK's for overall activities of KVKs.
2. **Total working experience:** It is total years of experience of working on the different cadres of TOs.
3. **Total training received:** It is total number of training received by TOs during service period in terms of short term training (up to 7 days), medium term training (8 to 15 days) and long term training (above 15 days), excluding M. Sc. / Ph. D. study.

3.3.1.3 Organizational Variables

1. **Span of control:** The span of control is known as span of management. It is a human resources management term that refers to the number of subordinates a training organizer had to manage.
2. **Availability of technical staff:** It refers to total number of technical personnel assisting to TOs in any KVKs to run overall

activities of KVK. It was measured as per the method used by Patel (1990).

3. **Organizational climate:** It is overall atmosphere of KVKs as experienced by TOs. It was measured in terms of healthy, good, average and adverse. It was measured as method developed by Talukadar, 1984.
4. **Organizational facilities:** It refers to physical facilities possessed by KVKs in terms of vehicle, audio-visual aids and other facilities. It was measured as method used by Patel (1993).
5. **Inter-personal communication:** It refers to degree or frequency of exchange of ideas or any information made by TOs with the higher and lower staff member. It was measured in terms of fortnight, monthly and more then monthly. It was measured by scale used by Popat (1991).

3.3.1.4 Socio-Psychological Variables

1. **Personality of to:** It refers to introvert and extrovert qualities of TO, it was measured by scale developed by Woodworth (1979).
2. **Attitude towards extension work:** It refers to positive or negative feeling of TO towards extension work. It was measured by scale developed by Daftur (1986).
3. **Workload:** it refers to total workload possessed by TOs as perceived by them. It was measured in terms of very heavy, heavy, moderate and low. It was measured as method used by Patel (1990).
4. **Job involvement:** It refers to self-involvement of TO in administrative, extension, research and all other activities of KVKs. It was measured in terms of full, partial, occasional and no involvement in above activities it was measured as method developed by Lodalh and Kejner (1965).
5. **Job stress:** It refers degree of pressure and tension of work as perceived by TO during working period. It was measured in terms of the perception of TOs *viz.*, very high, medium and low. It was measured as method developed by Talukadar (1984).

6. **Constraints:** It refers to items of difficulties faced by the TOs in performing their activities.

3.4 Derivation of Hypotheses

Based on the theoretical framework and keeping objectives in view statement and importance of study, the following statistical hypotheses (null form : H_0) were formulated as per the procedure recommended by Kerlinger (1976).

H_{01}: There will not be any relationship between personal, job oriented, socio-psychological and organizational character of training organizers of KVKs and their managerial ability.

H_{02}: There will not be functional relation between managerial ability and various independent variables.

H_{03}: There will not be any constraints faced by training organizers of KVKs in managing KVK activities.

3.5 The Conceptual Model

In the light of the above theoretical frame and the assumption derived there upon, a conceptual model delineating the relationship between independent variables and dependent variable has been proposed. The variables namely; age, academic qualification, family size, health, type of family, size of family, native place, total experience in KVK, experience as KVK head, total training received, span of control, interpersonal communication, organizational climate, infrastructure facility, technical staff facility, personality, perceived workload, attitude towards extension work, job involvement and job stress were taken as independent variables. The extension management ability was considered as dependent variable. Based on the above variables, the conceptual model and paradigm model were developed.

4. RESEARCH METHODOLOGY

This chapter deals with research design, tools and techniques of a scientific investigation used for data collection in the light of objectives of the study. The selection of universe and sampling technique for investigation as well as devices used for data analysis are also explained in this chapter under following sub heads.

4.1 Plan of study

4.2 Local of the study

4.3 Sample size and selection of the respondents

4.4 Research Design
4.5 Tools for the study
4.6 Pre-testing of the Personal interview schedule cum questionnaire
4.7 Method of data collection
4.8 Selection of the variables
4.9 Measurement of dependent variable
4.10 Measurement of independent variables
4.11 Statistical framework for analysis of data
4.12 Derivation of research hypothesis

4.1 Plan of Study

The Krishi Vigyan Kendra (KVK) is an important agricultural extension project of the Indian Council of Agricultural Research (ICAR) to test and transfer agricultural technologies to bridge the gap between the production and productivity and to increase self-employment opportunities among farming communities. The training organizer as a head of KVK has to perform the role of extension manager to get the maximum output from the available resources. The management ability of training organizer is important in achieving the maximum extension output for any KVK. To appreciate potential of KVKs, real understanding of the managerial ability of the training organizers is needed. However, the study on extension management ability of training organizers of KVKs was lacking. Only few studies have been conducted on role performance/productivity/effectiveness of academicians, research personnel and extension workers. Hence, it was found worth to conduct a study on such an important aspect.

The idea was then discussed with research guide, teacher counselors and members of advisory committee as well as extension personnel. FInally, it was considered that the study of this nature would be fruitful and would provide new direction in the field of agricultural extension management. The outline of research work (ORW) was presented against panel of experts appointed by Yashwantrao Chavan Maharashtra Open University, Nashik, Maharashtra and got it approved for research investigation. This way it was decided to undertake the study on “Extension Management Ability of Training Organizers of Krishi Vigyan Kendras”.

4.2 Local of the Study

The present study was carried out in all eight zones of Krishi Vigyan Kendras of India, where Krishi Vigyan Kendras are in operation in transfer of technology process. The ICAR is the apex body to conduct agricultural research and extension activities at the national level. KVK is one of the schemes of the council designed to provide skill oriented vocational training to the farmers, farmwomen, rural youths and field functionaries. The organizational structure has been suggested and developed by the council to maintain the scheme. It was therefore, decided to encompass the study on training organizers of about half of those KVKs which were in operation with full-fledged activities in the service of Indian farmers through effective agriculture extension services atleast with five years. At the time of investigation, about 320 well established KVKs were short listed. The KVKs in India are divided in eight zones for easy administration, thus purposively the representation of each zone was included in this study. All zones are headed by separate zonal coordinators at their Head Quarters/Zonal Coordinating Unit Office. Zonal coordinators are in charge for monitoring, evaluating and controlling the KVKs of their concerned zones.

Thus, the present study was an attempt to evaluate and assess the extension management ability of the training organizers of KVKs of India, which is to some extent fresh work in this direction.

4.3 Sample Size and Selection of the Respondents

The present study was undertaken in all the eight zones of the country. The representation of all the states and all eight zones of India where, Krishi Vigyan Kendras are in operation in transfer of technology process was included for this study. A random sample of 12 to 20.

Training Organizers of those KVKs which have completed atleast 5 years of working were selected from all the eight zones. Thus a random sample of 160 Training Organizers of KVKs was selected for this study. Out of selected total 160 Training Organizers of KVKs from the 28 States of India, 85, 56, 17, 1 and 1 KVKs were managed by SAUs, NGOs, ICAR, Deemed University and Open University, respectively. As far as state wise selection of Training Organizers of KVKs was concerned, from the Andaman and Nicobar-1, Andhra Pradesh-10, Arunachal Pradesh-1, Assam-4, Bihar-11, Chhattishgarh-6, Gujarat-12, Haryana-8, Himachal Pradesh-5, Jammu and Kashmir-1, Jharkhand-4, Karnataka-8, Kerala-

10, Lakshdweep-1, Madhya Pradesh-6, Maharashtra-10, Manipur-1, Meghalay-1, Mizoram-1, Nagaland-1, Orissa-10, Punjab-6, Rajasthan-11, Sikkim-1, Tamilnadu-8, Tripura-1, Uttar Pradesh-10, Uttaranchal-6 and from the West Bengal 5 Training Organizers of KVKs were selected for the study.

4.4 Research Design

As the study is concerned with the understanding the managerial ability of the Training Organizers or head of the KVKs, Ex-Post-Facto Research Design was used for this study. Kerlinger (1976) stated that the Ex-Post-Facto Research Design is worthy to apply when the independent variables have already acted upon. Ex-post facto research is systematic empirical investigation in which, the scientists do not have direct control on independent variables because they are inherently not manipulatable in nature. In this kind of research, researcher's control on the independent variable is very weak and in many cases no control is possible. Social science dose not afford a possibility of controlling the changes in the behavioural pattern of independent variables. These are usually affected by complex social phenomenon and one can only examine how the dependent variable is affected by such situations. In social and scientific research, there always exists a possibility of identifying dependent and independent variables either singly or in groups. Ex-Post-Facto Research is a good explanatory instrument of changes that take place in the dependent variables. If the set of independent variables is carefully selected and proper techniques adopted for analysis, meaningful insight can be gained into changing pattern of dependent variables.

4.5 Tools for the Study

The data were collected by personal interview as well as by mail questionnaire. Personal interview schedule cum questionnaire has been considered to be the most important tool and researcher can get most authentic first hand information. The interview schedule cum questionnaire was prepared by keeping in view the objectives of the study and was common for all the respondents. In formulating the questions and statements for the schedule, the investigator took opinion and guidance of the research guide, experts of the Department of Extension Education and Agricultural Statistics of B. A. College of

Agriculture, Anand Agricultural University, Anand as well as experts of YCMOU, Nasik, Maharashtra.

The interview schedule cum questionnaire was divided into five parts. First part was containing some questions to collect personal and economic information, second part contained organizational information of KVKs. The third part was of socio-psychological information of training organizers. The fourth part was in relation to extension management ability of training organizers along with ten main indicators and 50 sub indicators of all main indicators. The fifth and final part of the interview schedule cum questionnaire was regarding the constraints faced by the training organizers and their suggestions to over come their constraints.

4.6 Pre-testing of the Personal Interview Schedule cum Questionnaire

In order to test the administration ability of each item, question and statement and to remove ambiguity in measuring instrument, pre-testing of personal interview schedule cum questionnaire was carried out with 10 non-sample Training Organizers of zone 6, which comprises Gujarat and Rajasthan States. The training organizers were randomly selected from both of the states. The respondents were informed about the importance and purpose of the study. After gaining their confidence and developing rapport with them interview schedule was put forth to them. On the basis of answers given by these respondents and hurdles faced during interviewing, necessary changes were made after consultation with the research guide. The final approved interview schedule has been given in appendices.

4.7 Method of Data Collection

The data were collected by personal interview as well as mail questionnaire. The personal interview was carried out at the time of National Gathering of Krishi Vigyan Kendra of India at Rajasthan Agricultural University, Bikaner during 18-20 September-2005. Some of the responses were also collected from some of the training organizers by mail. While collecting data through personal interview, at the outset, the researcher introduced himself to the selected training organizers. The objectives of the study were explained with a view to facilitating correct responses. Assurance was given that the data collected would be utilized for the purpose of the study only. After establishing rapports, questions were asked one by one and their responses were recorded. Every possible cares were taken to maintain a favourable and unbiased opinion of the respondents. Similarly, while collecting responses through mail questionnaire; necessary cares were taken to make all the matters clear and understandable through written words. Before sending mail questionnaire, some times researcher also tried to contact some of the training organizers through telephone to clear the matter and purpose as well as to get better responses.

4.8 Selection of the Variables

The variables under study have been selected on the basis of extensive review of literature related to the subject and consultations with experts are as under. The major objective of this study was to measure the extension management ability of the training organizers of KVKs. Thus, the extension management ability of training organizer of KVKs was considered as dependent variable and overall characteristics of the training organizers were considered as independent variables.

4.8.1 Dependent Variable

1. Extension management ability.

4.8.2 Independent Variables

Personal Variable

1. Age
2. Educational qualification
3. Health status
4. Type of family
5. Size of family
6. Native place

Job related Variables

1. Experience as a training organizer of KVKs
2. Total working experience in KVK
3. Total training received.

Organizational Variables

1. Span of control
2. Availability of technical staff
3. Organizational climate

4. Organizational facilities
5. Inter-personal communication

Socio-Psychological Variables

1. Personality of the training organizers
2. Attitude towards extension work
3. Work load
4. Job involvement
5. Job stress

4.9 Measurement of Dependent Variable

4.9.1 Extension Management Ability of Training Organizers of KVK

The term Management is defined by different scientists in different words. "It is the ability of a person in using the techniques and skills in planning, scheduling, guiding, supervising and organizing the resources, monitoring and controlling of a research / development or any other projects" (Venkateshwarlu, 1988). In the present study extension management ability of the training organizers of KVKs has been operationalized as the ability of the head to apply the basic principles of management in running the KVKs. The extension management ability of the training organizers of KVK was measured by using the scale developed by Patel (2001) with slight modifications.

Many scientists have contributed to the meaning and concept of management in their own way. Finally one can define management as the effective and efficient integration and coordination of resources to achieve desired objectives.

The term, extension management can be defined as an effective and efficient utilization and coordination of available resources to achieve the objectives of extension organizations including Krishi Vigyan Kendra. The training organizer has to develop favourable coordination of various extension systems and also has to share the responsibility of extension education with other agencies and organizations. All this calls for new policy initiatives, creative management of the extension systems and all organized efforts to upgrade the knowledge and skills of extension functionaries. Thus, a training organizer has to perform a variety of tasks related to extension management.

To measure the extension management ability of the training organizers, the ten main indicators taken into consideration were; Planning, Organizing, Directing, Communicating, Controlling, Human relation, Leading, Supervising, Coordinating and Decision-making. The scale value for main and sub indicator were used as per the original scale. Total ten main indicators were measured by sub-indicators as seen in Appendix.

For measuring the extension management ability of training organizers of KVKs, the developed scale was applied to randomly selected respondents of selected KVKs. The responses were collected from the respondents on a three point continuum with categories *viz.*, 'Always', 'Sometimes', and 'Never' and the scores of 3, 2 and 1, were assigned, respectively. The formula used for calculating the extension management ability index was as under:

$$\text{EMAI} = \frac{\Sigma(\text{score obtained for sub item} \times \text{Weightage for sub item}) \times \text{Scale value of main item}}{\Sigma\,(\text{Maximum score for sub item} \times \text{Weightage for sub item}) \times \text{Scale value of main item}} \times 100$$

The extension management ability index of all training organizers was calculated item wise. The final extension management ability index of training organizer was determined by averaging the index of all 10 sub-items of the respective training organizers. Then, the training organizers were classified into three categories on the basis of Mean and Standard Deviation, viz., low (below mean – S.D.), medium (mean ± S.D.) and high (above mean + S.D.). Similarly, to classify training organizers in to three groups based on all the ten sub indicators also mean and standard deviation were used.

4.10 Measurement of Independent Variables

The details of measurement of independent variables are given below:

4.10.1 Age

Age of the training organizer was operationalized as number of completed years as on date of response. To quantify this variable one score to each year of age was given. The training organizers were classified into three groups, *viz.*, young (upto 35 years), middle (36 to 45 years) and old (above 45 years).

4.10.2 Educational Qualification

It refers to the training organizer's formal degree. The Training Organizers were classified into three groups as Doctoral degree holder (20 scores), Master's degree holder (17 score) and Bachelor's degree holder (15 score).

4.10.3 Health

The dictionary meaning of health is physical and mental well being, free from disease or a condition of body/mind. The health of training organizers was measured on the basis of their own perceptions. The training organizers were classified into five groups according to their self perception, *viz.*, very good health, good health, average health, poor health and very poor health and scores of 5, 4, 3, 2 and 1 were assigned to quantify this variable.

4.10.4 Type of Family

It refers to type of family of training organizers in terms of nuclear or joint type of family. The scores of 1 to nuclear type of family and 2 to joint type of family were assigned to compute this variable.

4.10.5 Size of the Family

Family size constitutes the total number of members of the family. One score was designed to each member of the family to quantify this variable one score to each member was given. The training organizers were grouped into three categories as small (below mean – S.D.), medium (mean ± S.D.) and big (above mean + S.D.).

4.10.6 Native Place

It refers to the birthplace of the training organizer. The training organizer were grouped into two groups *viz.*, "Rural" and "Urban" and scores of '2' and '1' were assigned, respectively to calculate this variable.

4.10.7 Experience as a Training Organizer of KVKs

It was operationalized as number of completed years of service as training organizer of KVK. The training organizers were classified into three categories *viz.*, low (below mean – S.D.), medium (mean ± S.D.) and high (above mean + S.D.) level of experience as training organizers of KVKs. One score was assigned to each year of experience to quantify this variable.

4.10.8 Total Working Experience in KVK

It refers to total years of working experience of training organizer on the different cadres in KVKs. It means length of service in KVK from the date of joining the service to the date of response was taken as total experience in KVK. One score was assigned to each year of experience to measure this variable. The training organizers were classified into three categories as low (below mean – S.D.), medium (mean ± S.D.) and high (above mean + S.D.) to working experience in KVK.

4.10.9 *Total Training Received*

Total training refers to the training received excluding M.Sc./ Ph.D. study by training organizers during service period. The training organizers were classified into two categories viz., low (below mean) and high (above mean) trained. The scoring technique used to quantify this variable is given in appendix.

4.10.10 *Span of Control*

It is the total number of persons presently working in KVK under training organizer. The training organizers were classified in three categories viz., low (below mean – S.D.), medium (mean ± S.D.) and high (above mean + S.D.). The scoring technique used to quantify this variable is given in appendix.

4.10.11 *Availability of Technical Staff*

It refers to total number of technical personals assisting to training organizers in any KVKs to run overall activities of KVK. It was measured as per the method used by Patel (1990). It was measured on the three point continuum *viz.*, "more than requirement" (Excess), "As per requirement" (sufficient) and "less than requirement" (insufficient) and scores of 2, 3 and 1 were given, respectively. The same scoring method was used for each cadre *viz.*, technical, administrative and auxiliary staff. The training organizers were classified into three categories as low (below mean – S.D.), medium (mean ± S.D.) and high (above mean + S.D.).

4.10.12 *Organizational Climate*

It is overall atmosphere of KVK as experienced by training organizers. It was measured by using scale developed by Talukadar (1984). The responses were collected in terms of highly favourable, quite, some what and not at all favourable and the scores of 4, 3, 2 and 1, were assigned respectively to measure this variable. The training organizers were grouped in to three categories *viz.*, low (below mean – S.D.), medium (mean ± S.D.) and high (above mean + S.D.).

4.10.13 Organizational Facilities

It refers to physical facilities possessed by KVKs in terms of vehicle, audiovisual aids and other infrastructural facilities for effective extension work of KVK. This variable was measured with the help of a developed questionnaire and responses were collected for each item in four point continuum viz., available as per requirement, can be easily available from other, can available with some difficultly and can not be available and scores of 4, 3, 2, and 1 were assigned, respectively to measure this variable. The final score was received by summing scores of all the items. The training organizers were classified into three categories as low (below mean – S.D.), medium (mean ± S.D.) and high (above mean + S.D.).

4.10.14 Inter Personal Communication

It refers to degree or frequency of exchange of ideas or any information made by training organizers with the higher and lower staff members. It was measured by preparing questionnaire and responses were collected in five points continuum in terms of contacts of training organizer with supporting staff. The responses were collected in terms of very frequently (every day), frequently (once in two days), sometimes (once on week), seldom (once in month or more period) and never and scores of 5, 4, 3, 2 and 1 were assigned, respectively. The final score was received by summing scores all the items. The training organizers were categorized into three groups as low (below mean – S.D.), medium (mean ± S.D.) and high (above mean + S.D.).

4.10.15 Personality of Training Organizer

There are two types of personality viz., introvert and extrovert. The psychic energy of introvert training organizers is directed inward their thoughts, emotions and desires, whereas, the psychic energy of extrovert training organizer is directed outward to the social environment.

This variable was measured by adopting the scale developed by Woodworth (1979) with slight modification. The scale consisted of nine statements with a two point continuum *viz.*, 'yes and 'no' with score of 2 and 1 for positive and 1 and 2 for negative statements, respectively. The final score was obtained by summing scores of all the items. The training organizers were classified into two groups as introvert (up to Mean) and Extrovert (above mean).

4.10.16 Attitudes towards Extension Work

It refers to degree of positive or negative feeling of training organizer associated with delegation of authority in KVK. It was measured by implementing the scale developed by Daftur (1986). The scale consisted of ten statements with a five point continuum as strongly agree, agree, undecided, disagree and strongly disagree with score of 5, 4, 3, 2 and 1 for positive and 1, 2, 3, 4 and 5 for negative statements, respectively. The final score was received by summing scores all the items. The training organizers were classified into three categories as low (below mean – S.D.), medium (mean ± S.D.) and high (above mean + S.D.).

4.10.17 Perceived Workload

It refers to the workload perceived by training organizers for their own job. The training organizers were classified on the basis of their perception of workload into three groups *viz.*, Heavy workload (three score), sufficient workload (two score) and light workload (one score).

4.10.18 Job Involvement

It refers to the degree to which a training organizer is involved psychologically with his work or degree of the importance of work in his total self image. It was measured by using a scale developed by Lodahal and Kejner (1965). The scale consisted of 20 statements with a five point continuum viz. strongly agree, agree, undecided, disagree and strongly disagree with the scores of 5, 4, 3, 2 and 1 for positive statements and 1, 2, 3, 4 and 5 for negative statements, respectively. The final score was gained by summing scores of all the items. The training organizers were classified into three categories viz., low (below mean – S.D.), medium (mean ± S.D.) and high (above mean + S.D.).

4.10.19 Job Stress

It has been operationalized as the psychological and / or physical response state of the Training Organizer with regards to mental/physical stress upon him pertaining to his job. It was measured with the help of procedure adopted by Patel (2001). The responses were collected on a five point continuum *viz.*, strongly agree, agree, undecided, disagree and strongly disagree with the score of 5, 4, 3, 2 and 1 for positive and 1, 2, 3, 4 and 5 for negative statements, respectively. The final score was received by summing scores of all the items. The trainings organizers were grouped in to three categories as low (below mean – S.D.), medium (mean ± S.D.) and high (above mean + S.D.).

4.10.20 Constraints Faced by Training Organizers in Managing KVK and Suggestions Offered by them to Overcome Constraints

The constraints were obtained from all the respondent training organizers in managing KVK. To know their degree of importance,respondents were asked to give their responses in three point continuum i.e. very important, important and not important. The scores assigned were, two, one and zero for very important, important and not important responses, respectively. Finally, the mean score for each constraint was calculated for ranking them in terms of its importance.

The same procedure was applied to identify and quantify suggestions.

4.11 Statistical Framework for Analysis of Data

The data collected were analyzed, classified and tabulated in order to make the findings meaningful for interpretation and drawing conclusions. For this, different statistical methods were used. Source of data was subjected to analyze in terms of percentage and frequency whenever necessary. At the other places, mean score and standard deviation were calculated. To find out the correlation between selected independent variables and extension management ability, the coefficient of correlation was computed. To predict the extension management ability by personal, job related, organization and socio-psychological variables, the stepwise regression analysis was done. The details regarding statistical framework is presented in this part.

The following statistical methods were used in the present study.

4.11.1 Frequency and Percentage

Simple comparisons were made on the basis of frequency and percentage.

4.11.2 Arithmetic mean and Standard Deviation

These techniques were used for classification of respondents into different categories.

(i) The mean was obtained by dividing total score by the number of respondents.

$$\bar{X} = \frac{\Sigma xi}{n}$$

Where,

n = Number of observations

Xi = Total score

(ii) The standard deviation was obtained by the square root of the average of the square deviation from mean.

$$S.D. = \sqrt{\left[\frac{\Sigma (Xi - \bar{X})^2}{n-1}\right]}$$

Where,
x = General mean
Xi = Observed values
S.D. = Standard deviation
n = Number of observations

4.11.3 Coefficient of Correlation

Coefficient of correlation was computed to find out the relationship between the independent variables and the dependent variable employing following formula. The Pearson's product moment method suggested by Chandel (1974) for computing correlation coefficient (r) was used to study the relationship between independent variables and dependent variable.

$$r = \frac{\Sigma xy - \frac{\Sigma x \Sigma y}{n}}{\sqrt{\left[\Sigma x^2 - \frac{(\Sigma x)^2}{n} \quad \Sigma y^2 - \frac{(\Sigma y)^2}{n}\right]}}$$

Where,
r = correlation coefficient
X and Y are two variables
n = Number of observations
X = Independent variable
Y = Dependent variable
Σxy = Sum of product of the deviation of 'x' and 'y' from their mean
ΣX^2 = Sum of square of deviation of 'x' from their mean
ΣY^2 = Sum of square of deviation of 'y' from their mean

4.11.4 Stepwise Regression Analysis

The stepwise regression analysis was employed to predict the extension management ability by personal, job related, organizational and socio-psychological variables.

In the stepwise method, the regression analysis was started with regression of Y and X_1.................X_k taking singly. The variable giving the greatest reduction in sum of squares of 'Y' was first selected. Then the bivariate regression in which X1 appeared was worked out. The variate which gave the greatest additional reduction in sum of square after fitting X_1 was selected. All trivariate regressions that include both X_1 and X_2 were computed the analysis was continued till the last variate of which additional contribution was the least of all variates.

The prediction of equation was................

$$Y = a + b_1x_1 + b_2x_2 + b_3x_3 + ...+ b_kx_k$$

Where,

a = intercept

b = partial regression coefficient

x_1 ....x_k = Number of independent variables

After fitting the regression equation the 't' value for partial regression coefficients was tested for their significance.

4.11.5 Standard Partial Regression Coefficient

In order to assign the rank to various selected impendent variable the standard partial regression coefficient was used. To calculate the same the following formula suggested by Sendecor and Cochran (1967) was used.

$$b' = \frac{\text{S.D. of individual item}}{\text{S.D. of total items}} \times \text{Partial 'b'}$$

Where,

b' = standard partial regression coefficient

4.12 Derivation of Research Hypotheses

Keeping the objectives of the study in view the research hypothesis was formulated on different aspects of the study. While formulating the hypothesis nature or relationship between variables were determined on the basis of review of literature. The hypotheses were set up and presented in null form (H_0).

H_{01}: There will not be any relationship between personal, job oriented, socio-psychological and organizational character of training organizers of KVKs and their managerial ability.

H_{02}: There will not be functional relation between managerial ability and various independent variables.

H_{03} : There will not be any constraints faced by training organizers of KVKs in managing KVK activities.

5. FINDINGS AND DISCUSSIONS

The information pertinent to this study was collected from the training organizers through the interview schedules and questionnaires. The data thus, collected were classified, tabulated and analyzed in the light of the objectives of the study. The facts and findings derived after analyzing the information have been presented under the following main heads and discussed in succeeding pages:

5.1 Personal, job related, organizational and socio-psychological characteristics of the respondent training organizers.

5.2 Extension management ability of the respondent training organizers.

5.3 Relationship between Personal, job related, organizational and socio-psychological variables of the respondent training organizers and their extension management ability.

5.4 The functional relationship between extension management ability and independent variables.

5.5 Constraints faced by training organizers in managing KVK and suggestions offered by training organizers to overcome the constraints.

5.1 Personal, Job related, Organizational and Socio-psychological Characteristics of the Respondent Training Organizers

The extension management ability is influenced by different characteristics of the training organizers. It was beyond the scope of the present study to include all the characteristics of training organizers however, some important characteristics were selected and the findings have been presented in succeeding pages. The characteristics have been classified into four groups' *viz.*, personal, job related, organizational and socio-psychological.

5.1.1 Personal Characteristics

5.1.1.1 Age

Physical and psychological development of an individual is related to hisage. It influences the interest and needs of an individual.Thus, age plays a crucial role in the behaviour of every individual. The information collected regarding the age of the respondents were grouped as (i) Young age group (up to 35 years); (ii) Middle age group (35 to 45 years) and (iii) Old age group (above 45 years). The data in respect of age of the respondents are presented in Table 1.

Table 1 : Distribution of the training organizers according to their age

N = 160

Sr. No.	*Age group*	*Number*	*Per cent*
1	Young age group	15	9.37
2	Middle age group	51	31.88
3	Old age group	94	58.75
Total		160	100.00

It is evident from the data presented in Table 1 that majority (58.75 per cent) of the training organizers were found in old age group, followed by 31.88 per cent with middle age group while only 9.37 per cent of the respondents were from young age group.

From the above result, it can be concluded that majority of the training organizers were from middle to old age group. The management of KVK is skillful work; person involved in such work has to deal with many people and situations. In addition to this, he has to do multifarious activities. To carry out such work effectively, certain skill, talent, expertise, proficiency and abilities are needed in any individuals. In such a situation handling of organization might have given to experienced persons. This might be the probable reason to have middle to old age among the majority of the training organizers of KVKs. Similar finding was reported by Patel (2001).

5.1.1.2 Educational Qualification

Education is generally considered as the process of producing desirable changes in human behaviour in terms of knowledge, skill and attitude. Education is valued as means of increasing knowledge and information. It facilitates learning. An individual's behaviour is influenced by his formal education. It leads him towards mental development and shapes view points. Keeping this in view, the level of education of the training organizers of KVKs was studied. The training organizers were categorized into three groups on the basis of educational qualifications such as Graduate, Postgraduate and Doctorate level. The data regarding educational qualification of training organizers are presented in Table 2.

The Table 2 gives the idea that majority (71.88 per cent) of the training organizers were educated up to doctorate level, followed by 19.37 per cent with master degree, while few of them (8.75 per cent) were educated up to bachelor degree only.

Table 2: Distribution of the respondents according to their education

N = 160

Sr. No.	*Category*	*Number*	*Per cent*
1	Graduate degree	14	8.75
2	Postgraduate degree	31	19.37
3	Doctorate degree	115	71.88
Total		160	100.00

The result indicates that majority (71.88 per cent) of the training organizers working at KVKs were found with doctorate degree. As discussed earlier, person handling KVK needs ability to deal with many situations, activities and problems. For this reason, highly qualified persons might be preferred and appointed by the authority to run KVKs to achieve desired goals. This might be the reason to have doctorate degree among majority of the training organizers working at KVKs.

This finding was similar to the findings of Rani (1985), Samanta (1985), Padheria (1989) and Patel (2001).

5.1.1.3 Health Status

The dictionary meaning of health is physical and mental well being, free from disease. It is also defined as condition of body and/or mind. It is believed that healthy person can do any work very effectively. Strong physical and mental health makes person able to think and do any work with his fullest involvement. Persons involved in any managerial work needs to have powerful physical and mental status. To understand effect of health on the extension management ability of training organizers, data were collected and presented in Table 3. The information on the health was collected and training organizers were classified into five categories according to self perception *viz.*, Very good health, Good health, Average health, Poor health and Very poor health.

The data presented in Table 3 reflect that two fifth (40.00 per cent) of the training organizers realized that their health was good, followed by 36.25 per cent and 21.25 per cent of the training organizers perceived that they were having average health and very good health, respectively. Equal percentages (1.25 per cent) of the training organizers felt that status of their health was poor or very poor.

Table 3: Distribution of the respondents according to their health

N = 160

Sr. No.	*Health*	*Number*	*Per cent*
1	Very good health	34	21.25
2	Good health	64	40.00
3	Average health	58	36.25
4	Poor health	2	1.25
5	Very poor health	2	1.25
Total		160	100.00

It can be summarized from the above result that majority (61.25 per cent) of the training organizers felt that status of their health was either good or very good. As discussed in previous table that training organizers have to do many activities, while doing mental and physical work at centre and in village areas he needs to maintain his physical and mental health, thus training organizers might have tried to maintain both the health. In addition to this, training organizers have to keep themselves busy and active while doing all such work, it is said that busy and active person's fitness remains always energetic and healthy.

This might be probable reason to have good or very good level of perception about the health among majority of the training organizers working at KVKs.

The result of Patel *et al.* (1999) was in conformity with this finding.

5.1.1.4 Type of Family

The joint family system is character of Indian culture. In our country still joint family system is prevailing, at the same time nuclear families are also increasing day by day. It is said that joint family system helps an individual member to share the views and learn new knowledge, skill and attitude from the old aged members of the family. Simultaneously, it is assumed that because heads of nuclear family seem young or middle aged so there are more chances to take fast decision in nuclear family. Thus, type of family from which training organizer belongs to plays a big role on his behaviour. To understand consequence of training organizers' family background on their overall extension management ability, information was collected. As regards to the family type, the training organizers were classified into two categories as presented in Table 4.

Table 4 : The respondents according to their type of family

N = 160

Sr. No.	*Type of family*	*Number*	*Per cent*
1	Joint Family	32	20.00
2	Nuclear family	128	80.00
Total		160	100.00

The data presented in Table 4 indicate that majority (80.00 per cent) of the respondents had the nuclear type of family, while only 20.00 per cent of them had joint type of family.

Hence, it may be inferred from the above result that majority of the training organizers had nuclear type of family. The result presented in one of the subsequent tables says that majority of the training organizers were from urban background. It is observed that because of some limitations and needs urban people prefer to have nuclear family system. In addition to this, majority of the KVKs are located in rural or semi urban areas of the country. The training organizers working in KVKs have to stay near by the centers or in the premises of KVK; in this situation they might have preferred to live in nuclear family with wife and their children. This might be the reason to have nuclear family system, among majority of the training organizers.

5.1.1.5 Size of Family

The family size refers to the total number of persons live together in a family unit. It is also an important variable, which can play a role on the behaviour of an individual. Hence, the family size of the training organizers working in KVKs was studied and data are presented in the Table 5.

The data presented in Table 5 reveal that majority (60.00 per cent) of the training organizers had up to four members in their family, while 32.00 per cent of them had five to eight members in their family and only 8.00 per cent had above eight members in their family.

From the above result, it can be concluded that majority (60.00 per cent) of the training organizers had up to four members in their family.

Table 5: Distribution of the respondents by family size

N = 160

Sr. No.	*Family size*	*Number*	*Per cent*
1	Up to 4 members	96	60.00
2	5 to 8 members	50	32.00
3	Above 8 members	14	08.00
Total		160	100.00

The training organizers working in KVKs had nuclear type of family system; this might be the probable reason to have upto four family members among the majority of the training organizers of KVKs.

5.1.1.6 Native Place

The native place of training organizers of KVKs plays important role on their performance. It is assumed that persons, who have to work for the development of farmers should have knowledge about the life, social system, customs and psychology of the farmers. The training organizers of KVKs with rural background will have positive tendency towards farmers. To understand role of this factor on the performance of training organizers, information was collected and data with regards to the native place are presented in Table 6.

Table 6: Distribution of the respondents according to their native place

N = 160

Sr. No.	*Native place*	*Number*	*Per cent*
1	Rural area	52	32.5
2	Urban area	108	67.5
Total		160	100.00

The data presented in Table 6 reveal that majority (67.50 per cent) of the training organizers of KVKs had urban background, whereas, 32.50 per cent of the training organizers of KVKs were from the rural area. The above finding indicates that majority of the persons working as training organizers in KVKs were not from rural background. To understand feelings, needs, problems, mindset, mentality and customs of farmers in a better way, authority should motivate to those who have rural background to work as training organizer of KVK. In other words, the result also reflects that persons from rural background either did not get chance or show deprived interest in working with rural people.

5.1.2 Job related Variables

5.1.2.1 Experience as a Training Organizers in KVK

It is well known fact that man can learn from his own experiences. The experience has been considered as the best teacher. Man learns everyday new things from his day-to-day life and makes progress from trial and error through experiments. One can expect better performance on any work or job from those who have more years of experience to be a part of similar job or work. To understand effect of experience on the extension management ability on the performance of the training organizers of KVKs, information was collected and presented in Table 7.

Table 7: The respondents according to their experience of training organizer in KVK

N = 160

Sr. No.	*Total experience in KVK*	*Number*	*Per cent*
1	Low (below 3.98)	23	14.37
2	Medium (in between 3.99 to 5.02)	118	73.76
3	Wide (above 5.02)	19	11.87
	Total	160	100.00

The data seen in Table 7 reveal that majority (73.76 per cent) of the training organizers had medium level of total experience. The low and wide intensity of experience was observed among 14.37 per cent and 11.87 per cent of the training organizers, respectively. It is natural that establishment of KVK in India is in progressing stage and yet ICAR is trying to establish it in different states of India. It is therefore training organizers working in recently established KVK might have medium level of total experience in KVK.

The present finding was in contrast with the finding of Patel (2001).

5.1.2.2 Total Working Experience in KVK

The total experience of the training organizers indicates total year of working experience in the field of research, extension and teaching activities. Normally it is believed that varied experience of dissimilar activities, job or position gives opportunity to an individual to learn diversified activities, which leads person to improve his performance. Therefore, the total experience of training organizers in research, extension and teaching was worked out. With respect to the total service experience of the training organizers, the data are presented in Table 8.

The data presented in Table reveal that slightly more than half (51.88 per cent) of the training organizers had medium level of total experience in KVK on a different positions, whereas 25.00 per cent and 23.12 per cent of them had wide and low level of experience, respectively. As discussed in the previous table, it is obvious that founding of KVK in our country is in rising stage; however, ICAR is attempting to launch at least one KVK in each district. As a result of this, training organizers working in the KVKs under this study might have medium level of working experience as a training organizer of KVK.

The finding of this study was in contrast with the findings reported by Patel *et al.* (1998)* and Patel (2001).

Table 8: Distribution of the respondents according to their total experience in KVK

N = 160

Sr. No.	*Total experience*	*Number*	*Per cent*
1	Low (below 1.74)	37	23.12
2	Medium (in between 1.74 to 7.84)	83	51.88
3	Wide (above 7.84)	40	25.00
Total		160	100.00

Mean = 4.79 S.D. = 3.05

5.1.2.3 Training Received

As a brief review of terms, training involves an expert working with learners to convey certain areas of knowledge or skills to improve performance on current jobs. Development through training activities is a broad, ongoing multi-faceted set of activities to bring someone or an organization up to another threshold of performance, often to perform some job or new role in the future.

Training can be initiated for a variety of reasons for an employee or group of employees. It is preferred when a performance appraisal indicates to improve, to decide the standard of improvement, to expand an overall professional development programme, for succession planning to help an employee be eligible for a planned change in role in the organization, to test the operation of a new performance management system and to practice members of organization for specific subject matter.

Training improves the competency of the training organizers and keeps them informed on the latest developments. Training improves skill of trainee in doing his job efficiently. It also gives new techniques of work, orientation of forth coming practices. The training may be in-service or pre-service; it always enhances performance of individual in doing particular job. There are varieties of the new areas in which training organizers working at KVKs need to take training.

To expect best management from the training organizers working in KVKs, some of the important areas of training like; training on communications to increase diversity among them to understand a wide variety of languages and customs, training on computer skills for conducting administrative, office tasks and e-extension system, training on customer service to understand increased competition in today's global market place to meet the needs of their customers, training on diversity to know about how farmers and rural people have different perspectives and views to value diversity, training on ethics to know increasing expectations about new social responsibility and variety of values and morals of the workplace, training on human relations to decrease job stresses, misunderstandings and conflict, training on quality initiatives to acquaint skill of total quality management, training on safety to know dangerous while working with heavy equipment, hazardous chemicals and repetitive activities and training on modern innovations to understand latest trends, are needed to achieve best results and determined objectives of KVK. Understanding importance of training in improving performance of the training organizers, information regarding total training received by them was collected and data are depicted in Table 9.

Table 9: Distribution of the respondents according to training received by respondents

N = 160

Sr. No.	*Training received*	*Number*	*Per cent*
1	Low (below 11.06)	122	76.25
2	High (above 11.06)	38	23.75
Total		160	100.00

It is observed from Table 9 that more than three-fourth (76.25 per cent) of the training organizers received less training, whereas, slightly less than one fourth (23.75 per cent) of the training organizers received high training. It can be concluded from the above data that majority of the training organizers received low training. Accepting important role of training organizers in doing effective activity of their respective KVKs for the overall development of farmers and rural India, they are needed to be enough competent through training in above listed areas to guide farmers to cope up with present scenario of WTO, diversified interests and diversified farming system.

5.1.3 Organizational Variables

5.1.3.1 Span of Control

The span of control is also known as span of management. It is a human resources management term that refers to the number of subordinates a manager or supervisor can effectively manage. It is important concept particularly for the effective management of any organizations. When an organization manager's span of control becomes too large, it affacts the growth of his or her organization. Even the best managers tend to lose their effectiveness when they spend all their time managing people and their issues and are unable to focus on long-term plans and competitive positioning for the organization growth as a whole. It is important to note that all managers experience a decrease in effectiveness, when their span of control exceeds the optimal level.

The data accessible in Table 10 reveal that majority (80.62 per cent) of the training organizers had medium level of span of control, whereas, 13.75 per cent and 5.63 per cent of them had low and high level of span of control, respectively.

Table 10: The respondents according to their span of control

N = 160

Sr. No.	*Span of control*	*Number*	*Per cent*
1	Low (below 8.18 score)	22	13.75
2	Medium (in between 8.19 to 16.90 score)	129	80.62
3	High (above 16.90 score)	9	5.63
	Total	160	100.00

Mean = 12.54 S.D. = 4.36

From the above information, it can be concluded that majority (80.62 per cent) of the training organizers had medium level of span of control. This might be due to staffing pattern decided for KVKs by the ICAR to work effectively.

Similar findings were reported by Patel (1993) and Patel (2001).

5.1.3.2 Staff Facility

It refers to the availability of technical/ administrative / auxiliary staff as per requirement perceived by training organizers working in the KVK. With respect to staff facility perceived by training organizers, the information is presented in Table 11.

Table 11: The respondents according to their staff facility

N = 160

Sr. No.	*Staff facility*	*Number*	*Per cent*
1	Low (below 3.93)	38	23.75
2	Medium (in between 3.94 to 8.37 score)	80	50.00
3	As per requirement (above 8.37 score)	42	26.25
Total		160	100.00

Mean = 6.15 S.D. = 2.22

It can be observed from the above table that exactly half (50.00 per cent) of the training organizers opined that the strength of staff provided them was medium, whereas, 26.25 per cent of training organizers opined that existing staff strength was as per the requirement, whereas 23.75 per cent of them felt that insufficient staff was provided to their KVK.

From the above result, it can be concluded that majority (76.25 per cent) of the training organizers perceived that the strength of staff provided in the KVK was medium or as per the requirement. The probable reason for this might be that nowadays ICAR is taking keen interest to strengthen farmers' development activity and giving good attention in recruitment of staff to carry out various extension activities of KVKs effectively. This way ICAR has tried to make KVKs well equipped up to a considerable level to provide vocational training to the practicing farmers.

5.1.3.3 Organizational Climate

The organizational climate is existing state of any organization as to grow continuously, manage with problems contribute to positive attitude, a feeling of belongingness and self-respect of working members that facilitate or hinder the effective functioning of the system. It is the environment, which helps in assisting and interfering with growth and bottom-line success. The organizational climate influences critical employee behaviours such as communication, problem solving, accountability, quality and performance. If organization wants to perform effectively, it is necessary to identify the key areas which are hindering performance, reducing effectiveness and which might generate unexpected impression. Organization becomes successful because of the quality of work, employees perform. When employees are cared for and the right environment is created where there are no barriers to performance, their true value to the organization can be fully realized.

Table 12: The respondents according to their organizational climate

N = 160

Sr. No.	*Organizational climate*	*Number*	*Per cent*
1	Low favourable (below 38.62)	26	16.25
2	Medium favourable (in between 38.63 to 48.12)	111	69.38
3	High favourable (above 48.12)	23	14.37
	Total	160	100.00

Mean = 43.37 S.D. = 4.75

The KVK is such an important rural and farmer development oriented organization, where favourable climate is needed. To understand training organizers' feelings about how employees of KVKs feel about their jobs, their supervisors, their peers, top management and many other factors affecting their individual productivity and collectively the ability of the KVKs to achieve its objectives, information on organizational climate of KVK as perceived by training organizers was collected and presented in Table 12.

It is evident from Table 12 that majority (69.38 per cent) of the training organizers working in KVKs had perceived medium level of favourable organizational climate, while 16.25 per cent and 14.37 per cent of the training organizers had low and high level of favourable perception for their organizational climate. From this result it can be said that majority of the training organizers had medium level of feelings about consciousness of their staff members' realization about sensitivity of each others happiness, clarity of goal and objectives of

KVK, job satisfaction, confidence, trust, inter-personal discussions about the issues, degree of interest in taking initiatives and concern of doing work effectively.

This finding is in contrast to the finding reported by Patel (2001).

5.1.3.4 Organizational Facility

To achieve best result from any organizations, infrastructural facility plays a momentous role. KVK is such an organization, where a variety of works and activities have to carryout by the training organizers and his co-workers. The high-tech and employment generating training, publication and supply of literature, education and organization of demonstrations, e-extension, field visits, exhibitions, off campus training programmes, educational tour and celebration of special days to motivate rural farmers towards development are some of the key activities, which need many infrastructural facilities like land, laboratory, agricultural implements, vehicles, audio-visual aids, reading materials, furniture, stationary and other communication facilities like telephone, zerox machine, fax, computer and internet connections. It is proved that providing infrastructural facility is one of the key components to create workable and conducive atmosphere of any training, communication and academic organizations.

Table 13: The respondents according to their organizational facility

N = 160

Sr. No.	*Organizational facility*	*Number*	*Per cent*
1	Low (below 41.5)	08	05.00
2	Medium (In between 41.6 to 57.54)	119	74.38
3	High (above 57.54)	33	20.62
Total		160	100.00

Mean = 49.52 S.D. = 8.02

It is the role of facility management to ensure that everything is available for the other staff in an organization to do their work significantly. Facility may range from the small scale to the large scale facility. To understand status of infrastructure facility available at KVKs, information was collected and presented in Table 13.

The data in the same table reveal that majority (74.38 per cent) of the training organizers felt that existing organizational facility in their KVK was medium. On the other hand low and high level of organizational facility was perceived by 5.00 per cent and 20.62 per cent of the training organizers, respectively. It is important to note that ICAR is spending sizable budget to create all kinds of facility at KVKs.

5.1.3.5 *Interpersonal Communication*

Interpersonal communication is defined as communication that occurs between people who have known each other for some time. Importantly, these people view each other as unique individuals, not as people who are simply acting out social situations. There are many positive points of the interpersonal communication in organizations. The communication climate in any organization is a key determinant of its effectiveness. Interpersonal communication is important because of the functions it achieves. One reason persons working in any organization need to engage in interpersonal communication is so that they can gain knowledge about another individual. Social Penetration Theory says that we attempt to gain information about others so that we can interact with them more effectively. We can better predict how they will think, feel, and act if we know who they are. We gain this information passively by observing them actively, having others engage them or interactively and by engaging ourselves with them.

Organizations with supportive environments encourage worker participation, free and open exchange of information and constructive conflict resolution. In organizations with defensive climates, employees keep useful information to themselves, make only guarded statements and suffer from reduced morale. It is observed that employees are influenced by the communication climate in the organization. A supportive climate is having empathy, equality, spontaneity, problem orientation and description or a defensive climate is having evaluation, control, strategy, neutrality, superiority, and certainty.

In the overall development of any KVK, frequent contact of training organizer with their head, directors, officers of state agricultural university, state agricultural department and other, plays important role. Such contacts give opportunity to training organizers to discuss new problems, developmental issues, budget related matter, innovations, constraints and many other useful matters with other people.

Table 14: The respondents according to their interpersonal communication

N = 160

Sr. No.	Interpersonal communication	Number	Per cent
1	Low (below 22.25)	38	23.75
2	Medium (in between 22.26 to 28.89)	93	58.13
3	High (above 28.89)	29	18.12
Total		160	100.00

Mean = 25.57 S.D. = 3.32

Understanding the importance of interpersonal communication of training organizers for the effective management of KVK, data were collected and presented in Table 14. The data indicate that more than half (58.13 per cent) of the training organizers had medium level interpersonal communication, whereas, 23.75 per cent of the training organizers had low level of interpersonal communication. Only 18.12 per cent of the training organizers had high level of interpersonal communication.

From the above finding, it can be said that majority (81.88 per cent) of the training organizers had low to medium level of interpersonal communication. It is obvious that the training organizers have to do several activities to achieve mandates of KVK, thus, they might not have got enough time to contact repeatedly to the other officers. This might be the reason to have low to medium level of interpersonal communication among majority of the training organizers of KVKs.

5.1.4 *Socio-psychological Variables*

5.1.4.1 Personality of the training organizers

The personality, in psychology is the patterns of behaviour, thought and emotion unique to an individual and the ways they interact to help or hinder the adjustment of a person to other people and situations. The human mind could be divided into three significant components the id, the ego and the superego, which work together or come into conflict to shape personality. An individual's personality is developed through external stimuli. The psychologists have postulated five basic dimensions of personality; introversion-extroversion, friendly compliance–hostile noncompliance, will, neuroticism, and openness to experience.

Table 15: Distribution of the respondents according to their personality

N = 160

Sr. No.	*Personality*	*Number*	*Per cent*
1	Introvert (below 13.65)	37	23.12
2	Extrovert (above 13.65)	123	76.88
Total		160	100.00

The person who has to work to inspire many people towards positive direction should have mature personality, which is not dominated by infantile pleasure drives. He should bc characterized by a strong need to repeat certain acts or rituals. He should have enough confidence and feelings of social responsibility. The training organizers working at KVKs should have specific qualities to work effectively with rural people. Understanding this, personality in terms of likeness of training organizers to work with group, motivation, self confidence, friendly attitude, likeness to lead social gathering was thought to study, information regarding this is presented in Table 15.

It can be observed from the table 15 that majority of the training organizers exhibited extrovert personality. The probable reason for emerging as extrovert category might be that the training organizers in the present study have long experience in extension management as well as extension work. Thus, it can be said that majority of the training organizers were gregarious and unreserved persons. This kind of personalities have act, state or habit of being predominantly concerned with and obtaining gratification from what is outside the self. Such qualities are helpful to carryout effective training, communication and educational and motivational work. Thus, it was interesting to note that majority of the training organizers working in KVKs had interest to know in what is happening around them, open mind and often talkative, mentality to compare their own opinions with the opinions of others, likeness of action and initiative and nature to make new friends or new group, interest in new people.

Extrovert Personality helps person to do extension work very effectively because of combination of traits and attitudes which are directed away from oneself and concerned primarily with external reality and environment rather than with inner feelings and thoughts accompanied by highly sociable, active, assertive, energetic, enthusiastic outgoing, impulsive and emotionally expressive and talkative.

5.1.4.2 Attitudes towards Extension Work

Attitude is a way of thinking, acting, or feeling, manner, or behaviour of a person toward a situation or cause. It is a position of the body suggesting an action, purpose, emotion, or mental state. It is universally accepted fact that an attitude of an individual plays an important role in determining his behaviour. It is obvious that training organizers with positive feelings towards extension work will have more interest to do such work. It is said that one can get best result from any work when he loves that work. The training organizer with more mental, psychological, intellectual and intellectual attachment with extension activity will always lead his organizations towards grand success. Hence, the training organizers' attitude towards extension work was also included in the focus of present study. The distribution of the respondents according to their attitude towards extension work is presented in Table 16.

Table 16: The respondents according to attitudes towards extension work

N = 160

Sr. No.	*Attitudes towards extension work*	*Number*	*Per cent*
1	Less favourable (below 32.95)	34	21.50
2	Favourable (in between 32.95 to 45.25)	100	62.50
3	Highly favourable (above 45.25)	26	16.00
	Total	160	100.00

Mean = 39.10 S.D. = 6.15

It is evident from Table 16 that majority (62.50 per cent) of the training organizers had favourable attitudes towards extension work, followed by 21.50 per cent and 16.00 per cent of them had less favorable and highly favourable attitudes towards extension work, respectively.

It can be concluded that great majority (78.75 per cent) of the training organizers of KVKs had favourable to highly favourable attitudes towards extension work. The reason behind this might be that the training organizers must have joined KVK with the feelings to do better service of nation and farmers understanding the main aim of the KVK. It is incredible to make a note of such mental status among the majority of the training organizers of KVKs. Favourable attitude towards extension work among training organizers helps to create constructive, encouraging and sympathetic impression of KVK among rural farmers to connect themselves with KVK activities.

5.1.4.3 Workload

Workload is the amount of work carried by training organizers or assigned to a worker or position. The workload plays important role on the performance of any manager or organizers. Heavy workload effects performance of any manager in undesirable manner, on other hand light workload shows more than required number of human resource in organization. It is therefore, to carry out work of any organizers to achieve desired goal, desirable workload on each member or worker of any organizers is recommended. To understand real picture about the degree of workload of training organizers of KVKs, information was collected.

Table 17:The respondents according to their perception for workload

N = 160

Sr. No.	*Workload*	*Number*	*Per cent*
1	Light workload	1	00.62
2	Desirable workload	96	60.00
3	Heavy workload	63	39.38
Total		160	100.00

The data in Table 17 revealed that majority (60.00 per cent) of the training organizers had perceived desirable workload, whereas, 39.38 per cent of the training organizers perceived heavy workload.

Only 00.62 per cent of the training organizers had perception of light workload. Thus, It can be concluded that majority (60.00 per cent) of the training organizers of KVKs had perception of desirable level of workload. The result says that authority had tried to give desired level of workload to the training organizers of KVKs in an effort to anticipate best performance from them.

This finding is in contrast to the findings reported by Patel (1993) and Patel (2001).

5.1.4.4 Job Involvement

The simple meaning of job involvement is the action of involving with job, profession, work or duty. In this study, job involvement of the training organizers was measured in terms of degree to which he is involved psychologically with his work or what he thinks about significance of work. It is obvious that, for excellent result at the end of any work, person who has to carry out that work should have rational, intellectual, logical and constructive mental status of being involved in work. The favourable feelings towards assigned duty, job, aspiration to do work, willingness to bear responsibility, concern to the job and likeness of job are some of the important mental preconditions enquired among the training organizers to lead KVK towards desired directions. Understanding importance of this, information on the job involvement of training organizers was collected. The data regarding this are depicted in Table 18.

The data in Table 18 indicate that majority (80.00 per cent) of the training organizers were reasonable, followed by 11.25 per cent with less and 8.75 per cent with high favourable towards their job.

Table 18: The respondents according to their job involvement

N = 160

Sr. No.	*Job involvement*	*Number*	*Per cent*
1	Less favourable (below 64.90)	18	11.25
2	Reasonable (in between 64.90 to 81.14)	128	80.00
3	Highly favourable (above 81.14)	14	8.75
Total		160	100.00

Mean = 64.90 S.D. = 81.14

It can be included that majority of the respondents exhibited reasonable level of job involvement. The probable reason for the reasonable level of job involvement may be that the training organizers in the present study possessed favourable level of attitudes towards extension work.

This finding is in contrast to the findings reported by Patel (1993) and Patel (2001).

5.1.4.5 Job Stress

The job stress has been operationalized as reactions of training organizer with regards to their psychological and/or physical stress pertaining to their job. A stressful situation occurs when imbalance is observed between demands imposed upon training organizers by the circumstances and their ability to cope with demands. It is clear that KVK is not only human resource development centre but many other administrative, research related and technical works are also expected functions of it. Thus, the person who is leading KVK should have physically and mentally powerful mindset to do work in any dam situation. He should be enough strong, prompt, tension free, matured, accommodating and able to do any kind of work effectively with whatever available workforce he has. Understanding importance of ability to deal with stress situation in the leader of KVKs, fact about level of job stress of training organizers was studied.

From the data presented in Table 19, it is observed that majority (73.12 per cent) of the training organizers experienced medium level of job stress, whereas, 13.75 per cent with low level of job stress and 13.13 per cent of them had perception of high level of job stress.

It can be inferred from the above results that majority of the training organizers had experienced medium level of job stress. The majority of the training organizers under study had four to five years of working experience, it is said that experience makes man perfect with job and same time it makes him able to tolerate various kinds of stress. An experienced person learns by trial and error to tackle different situations while working. This might be the reason to have low to medium level of perception about the job stress among majority of the training organizers.

Table 19:Distribution of the respondents according to their job stress

N = 160

Sr. No.	*Job stress*	*Number*	*Per cent*
1	Low (below 50.70)	22	13.75
2	Medium (In between 50.70 to 68.82)	117	73.12
3	High (above 68.82)	21	13.13
Total		160	100.00

Mean = 59.76 S.D. = 9.06

This finding is in line with the findings of Talukdar (1984) and Patel (2001).

5.2 Extension Management Ability of the Respondent Programme Coordinators

To measure the extension management ability of Training organizers, ten indicators were studied *viz.,* planning, organizing, directing, communicating, controlling, human relation, leading, supervising, coordinating and decision making. The responses were obtained in three point continuums *viz.,* "always", "some time" and "never". The mean and standard deviation were worked out and based on mean and S.D., the training organizers were categorized in three categories such as low (Mean – S.D.), medium (Mean + S.D.) and high (Mean + S.D.) in relation to all of the ten main indicators. The details were presented in following tables.

5.2.1 Ability of Planning

The Planning is one of the important indicators of extension management ability of the training organizers. It shows scheduling style implemented by training organizers to determine the objectives, training needs, process of transferring technologies, utilization of funds, extension facilities and monitoring and evaluation work. With a view to understanding ability of planning activity as one of the indicators of extension management ability of training organizers, information was collected and presented in Table 20.

Table 20: Distribution of the respondents according to ability of planning.

N = 160

Sr. No.	*Ability of planning*	*Number*	*Per cent*
1	Low	24	15.00
2	Medium	54	33.75
3	High	82	51.25
Total		160	100.00

Mean= 22.52 S.D. = 1.98

The data presented in Table indicated that in case of planning, slightly more than half (51.25 per cent) of the training organizers had high level of planning ability, followed by 33.75 per cent and 15.00 per cent of them with medium and low level of planning ability, respectively. It is concluded that (85.00 per cent) of the training organizer found with medium to high level of planning ability. Proper guideline, systematic direction and continous supervision by concerned officers and ICAR might have made them able with medium to high level of planning ability.

5.2.2 *Ability of Organizing*

The ability to organize different functions is one of the important indicators of extension management ability requires in any training organizers to arrange various activities on a regular basis. To call regular meeting to provide work opportunity according to specializations and to invite suggestions from subordinates as well as to divide activity into groups and to delegate authority according to responsibility are important functions which require proper ability to organize functions. With a view to understanding this ability as one of the indicators of extension management ability of training organizers, information was collected and presented in Table 21.

The data in Table indicate that an equal number of (42.50 per cent) of the training organizers had medium and high level of organizing ability, followed by only 15.00 per cent of them had low level of organizing capacities. It can be clearly stated that majority (85.00 per cent) of the training organizers had medium to high level of organizing capacity in relation to extension management in their respective KVKs. The probable reason might be that training organizers have to carryout series of activities throughout the year; as a result of this they might have understood the importance to systematize various activities for their effective implementations.

Table 21: The respondents according to ability of organizing

N = 160

Sr. No.	*Ability of organizing*	*Number*	*Per cent*
1	Low	24	15.00
2	Medium	68	42.50
3	High	68	42.50
Total		160	100.00

Mean = 16.70 S.D. = 1.54

5.2.3 *Ability of Directing*

This variable was measured by knowing degree of frequency of training organizers in directing their subordinates about different functions. This is an important activity; training organizers have to carry out for effective functioning of the KVK work. In order to know this ability as one of the indicators of extension management ability of training organizers, information was collected and presented in Table 22.

Table 22: The respondents according to ability of directing

N = 160

Sr. No.	*Ability of directing*	*Number*	*Per cent*
1	Low	15	09.38
2	Medium	22	13.75
3	High	123	76.87
	Total	160	100.00

Mean = 11.65 S.D. = 0.75

It can be seen that majority of the (76.87 per cent) training organizers had high level of ability of directing their subordinates, followed by 13.75 and 9.38 per cent of them were with medium and low level of ability of directing their subordinates for efficient extension management activities of the KVKs. The results indicates that majority of the training organizers took high level of interest to give clear instruction of work, to give due credit to a person, to appreciate good work done by their subordinates and to give continuous information.

5.2.4 *Ability of Communicating*

The ability of communicating subordinates is also an important indicator, which requires in training organizers for effective extension management. Understanding importance of this ability as one of the indicators of extension management of training organizers, information was collected and presented in Table 23.

Table 23: The respondents according to ability of communicating

N = 160

Sr. No.	*Ability of communicating*	*Number*	*Per cent*
1	Low	34	21.25
2	Medium	33	20.62
3	High	93	58.13
	Total	160	100.00

Mean = 11.31 S.D. = 1.15

It is seen from the Table that majority of the training organizers (58.13 per cent) had high level of ability of communicating, followed by 21.25 per cent of them had low and 20.62 per cent of them had medium level of ability of communicating, respectively. Thus, it can be concluded from the result that majority of the training organizers had high level of ability of communicating. The result indicates that majority of the training organizers had taken constant interest to give timely instructions, to listen subordinates, to select proper medium for instructions and to receive feed back

5.2.5 Ability of Controlling

The ability of controlling is also an important sign, which requires in training organizers for effective extension management. This variable was measured by knowing degree of frequency of training organizers' command of ling their subordinates and different functions.

The data presented in Table stated that slightly less than half (48.12 per cent) of the training organizers had high level of controlling power, followed by 28.12 per cent and 23.76 per cent of them had low and medium level of controlling power, respectively. It is seen that majority (71.88 per cent) of the training organizer realized medium to high level of controlling power for effective functioning of extension activities in terms of monitoring the work, accurate and proper style of visits, determining standard for work, evaluating the work and measuring performance of staff.

Table 24: The respondents according to ability of controlling

N = 160

Sr. No.	*Ability of controlling*	*Number*	*Per cent*
1	Low	45	28.12
2	Medium	38	23.76
3	High	77	48.12
Total		160	100.00

Mean = 13.63 S.D. = 1.51

5.2.6 Ability to Maintain Human Relation

The ability to maintain human relation is very important quality; training organizers need to maintain effective extension management. This variable was measured by knowing degree of frequency of training organizers' touch with persons to carry out functions of KVK effectively. Accepting importance of this ability as one of the indicators of extension management in training organizers, information was collected and presented in Table 25.

Table 25: The respondents as per ability to maintain human relation

N = 160

Sr. No.	*Ability of human relation*	*Number*	*Per cent*
1	Low	13	08.13
2	Medium	70	43.75
3	High	77	48.12
Total		160	100.00

Mean = 17.01 S.D. = 1.32

The data revealed that great majority (91.87per cent) of the training organizers had medium to high level of capacity to maintain human relation, followed by only 8.13 per cent of them had low level of human relation making capacity. It is obvious from the table that majority of the training organizers had medium to high level of interest to make personal contacts, deal fairly, pay attention in development of juniors, adopt participative approach, understand others' behaviour and to offer good co-operation.

5.2.7 Ability of Leading

The ability of leading is significant feature; training organizers need to uphold for effective extension management. This variable was measured by knowing degree of training organizers' lead taking behaviour to perform the functions of KVK successfully. Considering importance of this aptitude as one of the indicators of extension management among training organizers, information was collected and presented in Table 26.

Table 26: The respondents as per their ability of leading

N – 160

Sr. No.	*Ability of leading*	*Number*	*Per cent*
1	Low	29	18.12
2	Medium	32	20.00
3	High	99	61.88
	Total	160	100

Mean = 14.29 S.D. = 1.16

The perusal of the data presented in the table indicates that majority (61.88 per cent) of the training organizers had high level of feeling of lead taking behaviour, followed by 20.00 per cent and 18.12 per cent of them had medium and low level of feeling of lead taking behaviour, respectively. It can be concluded from the results that great majority (81.88 per cent) of the training organizers realized that they had medium to high degree capacity to take lead in different functions like; to promote team work, take initiative task, apply technical competency, have a sense of responsibility and to inspire subordinates.

5.2.8 Ability of Supervising

The ability of supervising is another considerable trait; training organizers require for upholding effective extension management. This quality was measured by knowing degree of training organizers' administering behaviour to do various functions of KVK successfully. Considering essence of this talent as one of the indicators of extension management among training organizers, data were collected and presented in Table 27.

Table 27: The respondents as per their ability of supervising

N = 160

Sr. No.	*Ability of supervising*	*Number*	*Per cent*
1	Low	13	8.12
2	Medium	33	20.62
3	High	114	71.26
Total		160	100.00

Mean = 8.60 S.D. = 0.71

The data presented in Table 27 clearly indicate that majority (71.26 per cent) of the training organizers indicated high level of supervising capacities, followed by 20.62 per cent and 8.12 per cent of them had medium and low level of supervising capacities, respectively. Thus, it can be said that great majority (91.88 per cent) of the training organizer had medium to high level of ability to regulate various functions by providing constant extensive counseling, focusing on quality work and giving valuable suggestions. Due to frequent involvement in various kinds of activity by the training organizers of KVK, they might have established such ability in them.

5.2.9 Ability of Coordinating

The ability of coordinating is one more substantial attribute; training organizers require for keeping efficient extension management. This element was calculated by insightful degree of training organizers' behaviour to organize, synchronize, harmonize, match up and bring together various on-campus and off-campus functions of KVK fruitfully. Considering core of this talent as one of the indicators of extension management ability of training organizers, facts were collected and presented in Table 28.

Table 28: The respondents as per their ability of coordinating

N = 160

Sr. No.	*Ability of coordinating*	*Number*	*Per cent*
1	Low	43	26.87
2	Medium	17	10.63
3	High	100	62.50
Total		160	100.00

Mean = 11.21 S.D. = 1.15

It can be seen that majority of the training organizers were found with high level of coordinating ability (62.50 per cent), followed by 26.87 per cent of them with low and only 10.63 per cent of them with medium level of ability of coordination with concerned personnel in the mandatory working of the KVK. Thus, it was interesting to note that majority of the training organizers had high level of ability to establish linkages, integrate the extension work, establish mutual trust and get co-operation from outsides. The kind of works, training organizers have to do as a manager of KVK indicates that need to deal with internal and external personnel and agencies, while working with them, they might have realized importance to establish such ability. This might be the reason for above finding.

5.2.10 Ability of Decision Making

The ability of taking right decision at the right time is an added considerable characteristic; training organizers want for keeping competent extension management. This ingredient was studied by knowing degree of training organizers' behaviour to take different judgments to carry out various on-campus and off-campus activities of KVK at right time by proper ways. Taking into consideration, role of this capacity as one of the indicators of extension management ability of training organizers, information was collected and presented in Table 29.

Table 29: The respondents as per their ability of decision making

N = 160

Sr. No.	*Ability of decision making*	*Number*	*Per cent*
1	Low	23	14.38
2	Medium	65	40.62
3	High	72	45.00
Total		160	100.00

Mean = 13.99 S.D. = 1.33

The data presented in Table indicate that 45.00 per cent of the training organizers had high level of decision making capacity, followed by 40.62 per cent of them with medium and only 14.38 per cent of them with low level of decision making capacities, respectively. It can be stated that great majority (85.62 per cent) of the training organizers had medium to high level capacity to take timely decision, find out facts for decision, adopt participative decision style, taking unbiased decision for implementation.

5.2.11 Overall Extension Management Ability

The overall extension management ability of Training organizers was measured by combining all the indicators including planning, organizing, directing, communicating, controlling, human relation, leading, supervising, coordinating and decision making. The detail is presented in Table 30.

Table 30: The respondents according to extension management ability

N = 160

Sr. No.	*Extension management ability*	*Number*	*Per cent*
1	Low (below 73.33)	21	13.13
2	Medium (in between 73.33 to 86.13)	109	68.12
3	High (above 86.13)	30	18.75
Total		160	100.00

Mean = 79.73 S.D. = 6.40

It can be seen that majority (68.12 per cent) of the training organizers were observed with the medium level of overall extension management ability, while 18.75 per cent with high and 13.13 per cent with low level of overall extension management ability.

Thus, the extension management ability of the respondents was predominantly medium.

It can be concluded from the above finding that the extension management ability of training organizers of Krishi Vigyan Kendra was predominately medium. This might be due to constant involvement in so many activities, less exposure of training on management related aspects, medium level of attitude towards extension work and having medium level of infrastructural facility.

5.3 Relationship between Personal, Job Related, Organizational and Socio-psychological Variables of the Respondent Programme Coordinators and their Extension Management Ability

To ascertain the relationship between personal, job related, organizational and socio-psychological variables of the respondents and their extension management ability, correlation coefficient was worked out. On the basis of operational measures developed for the variables, null hypothesis were stated for testing the relationship and their level of significance are given Table 31.

Table 31: Relationship between the characteristics of the respondents and their extension management ability

N = 160

Sr. No.	*Name of variables*	*Correlation coefficient*	
I	Personal variables		
1	Age	-0.362	**
2	Educational qualification	0.238	*
3	Health	0.365	**
4	Type of family	-0.011	NS
5	Size of family	0.020	NS
6	Native place	0.384	**
II	Job related variables		
1	Total experience as TO	-0.060	NS
2	Total experience in KVK	-0.013	NS
3	Training received	-0.019	NS
III	Organizational variables		
1	Span of control	0.144	NS
2	Staff facility	-0.120	NS
3	Organizational climate	0.524	**
4	Organizational facility	0.537	**
5	Interpersonal communication	0.401	**

contd...

IV	Socio-psychological variables		
1	Personality	0.675	**
2	Attitudes towards extension work	0.622	**
3	Workload	0.023	NS
4	Job involvement	-0.047	NS
5	Job stress	-0.158	*

* Significant at 0.05 level of probability.

** Highly significant at 0.01 level of probability.

5.3.1 Age and Extension Management Ability

In respect of age, the data presented in Table 31 indicate that the correlation coefficient (r = –0.362) was negative and significant with extension management ability of training organizers. The result says that management ability was observed better among those training organizers who had young age. In other words young training organizers were observed superior in planning, organizing, directing, communicating, controlling, leading, supervising, controlling and decision making than old aged training organizers. The young training organizers due to their high level of enthusiasm to prove themselves as an important leader of organization might have taken keen interest in performing diversified activities of KVKs than old aged training organizers. This might be a probable reason for above finding. It is natural that the young persons are physically more powerful, active and able in using modern equipments and facilitates of communication like computer, internet and e-mail, prompt, conscious in using power, able to control difficult situations and interested in doing work in spite of risk involved than old aged person. This might be another reason for above findings.

From the above result, it can be concluded that age had negatively significant relationship with extension management ability of the training organizers.

Hence, the null hypothesis (H_{01}) in this case of age was rejected and it was concluded that there was negative but significant relationship between age and extension management ability of the training organizers.

This finding is in conformity with the findings of Rani (1985) Padheria (1989) and Patel (2001).

5.3.2 Educational Qualification and Extension Management Ability

A perusal of data presented in Table 31 indicates that extension management ability of respondents had positive and significant relationship with their educational qualification. It means highly qualified training organizers seemed to be better than the less qualified ones. The reason for this might be that highly educated training organizers might have greater reception power which resulting in to more knowledge acquisition. In addition to this higher educated training organizers might have utilized their modern knowledge received, while earning high level of degrees to acquire ability to do better management. Further, higher education opens the world of thoughts leading towards positive thinking, better planning and proper execution. Thus, the training organizers with higher education might have higher management ability to work as efficient extension workers.

Hence, the null hypothesis (H_{01}) in this case of education was rejected and it was concluded that there was positive and significant relationship between education and extension management ability of the training organizers.

This finding is further strengthened by the results reported by Sharma *et al.* (1988), Padheria (1989) and Patel (2001).

5.3.3 Health and Extension Management Ability

The data in Table 31 indicate that there was positive significant relationship between health of the training organizers and their extension management ability. It means extension management ability was observed better among those training organizers, who had vigorous status of health. It is natural that person having strong mental and physical status will be very positive and capable to carry out any kinds of mental, psychological, intellectual, rational and physical works. The management of any organization always needs strong will power, positive attitude, capacity to deal with any situation mentally and physically, among those who are directly involved in it. This is possible if person bears strong health. The strong health provides all kind of conducive, favourable, encouraging, helpful and contributing situations to the person for above said pre-requisites to do effective management. This might be the probable reason to have high level of management ability among those training organizers who had sturdy health.

From the above discussion, null hypothesis (H_{01}) in this case of health was rejected and it can be concluded that there was significant relationship between health of training organizers and their extension management ability.

5.3.4 Type of Family and Extension Management Ability

The 'r' value (–0.011) seen in the Table indicates negative and non-significant relationship between the type of family of training organizers and extension management ability of training organizers. The negative relationship indicates that training organizers with nuclear family had high level of extension management ability than those, who were from joint family. It is understandable that the training organizers from nuclear family have less number of family members to take care of. This situation might have given more chance to involve and think more for office work than those who were from joint family system. In fact this trend was not up to the level of significance thus, it can be said that family type had least role on the extension management ability of the training organizers.

Thus, the null hypothesis (H_{01}) was accepted and it was concluded that there was negative and non-significant relationship between type of family and extension management ability of training organizers.

5.3.5 Size of Family and Extension Management Ability

The number of members in the family of the training organizers did not play a vital role in improving extension management ability of training organizers. Thus, a non-significant relationship was inferred between the size of family and extension management ability of training organizers. It was seen that as the number of family member increased, extension management ability was decreased and vice versa but this tendency was non-significant.

Thus, the null hypothesis (H_{01}) in case of family member of the training organizers was accepted and concluded that there was non-significant relationship between size of family and extension management ability.

This finding is in conformity with the finding of Patel (1993).

5.3.6 Native Place and Extension Management Ability

The relationship between native place of the training organizers and their level of extension management ability was positive and significant. As the method adopted for scoring this variable says that training organizer from urban background was given one score while training organizer from rural background was given two scores.

The result clear-cut indicates that training organizers with rural native background were observed more active and commanding in managing extension activities of KVK.

As we know, majority of the Indian farmers live in rural area, thus person working as training organizers of KVK has to work with rural people. It is logical that the person who has rural background will have more knowledge about mentality, attitude, way of thinking, mindset, local language, customs, traditions, convention, habit, routine, tendency and problems of farming community. In addition to this, person with rural area knows and understands better about farming situation prevailing in rural area. Thus, person with such rural background can work effectively with rural people. Such kind of possibility is comparatively less in case of person who is working in rural area alongwith his urban background. This might be the probable reason to have high level of extension management ability among those training organizers who were from rural background.

Thus, the null hypothesis (H_{01}) was rejected and it was concluded that there was positive and significant relationship between native place and extension management ability of training organizers.

5.3.7 Total Experience as to and Extension Management Ability

The correlation coefficient between total experience of training organizers and their level of extension management ability was –0.060 indicating non-significant relationship between them. The negative value indicates that level of extension management ability of those training organizers was better who had less experience of service than those of more experienced training organizers. It can be said that training organizers who were less experienced were younger than experienced one. As discussed earlier that young persons due to their elevated enthusiasm, they give the impression of being active in understanding modern way of working and system. This might be the reason to have negative relationship between these two variables.

However, 'r' value was statistically non-significant hence, the null hypothesis, (H_{01}) was accepted and it can be concluded that total experience had non-significant relationship with extension management ability of the training organizers.

This finding is further strengthened by the results reported by Rani (1985), Sakaria (1988), Patel (1996) and Patel (2001).

5.3.8 Total Experience in KVK and Extension Management Ability

The correlation coefficient between total experience of training organizers working in KVK and their level of extension management ability was indicating non-significant relationship. The negative value indicates that level of extension management ability of those training organizers was observed better who had less experience of service in KVK than those of more experienced training organizers or it can be said that training organizers who were less experienced and newer than experienced one were better as extension managers. In other words, it can be said that new training organizers of KVK due to their prominent eagerness to be an important person of KVK gave the impression of being vigorous in proving themselves as active extension managers. This might be the reason to have negative relationship between these two variables. Nevertheless, this trend was not seen up to the considerable level.

Therefore, the null hypothesis (H_{01}) was accepted and it can be concluded that total experience of training organizers with KVK had non-significant relationship with their extension management ability.

This finding is in conformity with those findings reported by Patel (1998) and Patel (2001).

5.3.9 Training Received and Extension Management Ability

The data in Table point out that there was negative insignificant relationship between training received by training organizers and their extension management ability. In other words, it can be said that training received by the training organizers did not exhibit any impact in improving extension management ability of the training organizers. The training is said training if it shows any impact on the performance of any individual on his job. There is no meaning of any training, if it is not mental or physical skill improvement oriented. It is seen that many training organizers are failed to prove their ability in bringing desirable improvement on the performance of the trainees on their job. In such a situation, in spite of exposure of person under many training, one cannot expect better results. This was observed in the present study.

This may be due to the reason that the general training does not support much to management aspects. Thus, the training organizers with general training mostly on subject matters could not improve much their skills of managerial ability of extension work.

Hence, the null hypothesis (H_{01}) in case of training received was accepted and it was concluded that there was non-significant relationship between training received by the training organizers and their extension management ability.

Similar findings were observed by Rani (1985), Sakaria (1988), Patel *et al.*, (1998) and Patel (2001).

5.3.10 Span of Control and Extension Management Ability

The information available in Table designates that span of control has non-significant relationship with extension management ability of training organizers. Though the span of control has not played any significant role in increasing extension management ability, positive relationship indicates that extension management ability of those training organizers was better who had more workers to look after. In other words, it can be said that training organizers with more members to control directly by them were more active in managing KVKs. It also indicates that training organizer who had adopted strategy to maintain direct control and had more observation on all members of their KVKs had better extension management ability. However, this inclination was not up to the level of importance.

Thus, the null hypothesis (H_{01}) in case of span of control and extension management ability was accepted and it can be concluded that span of control had non-significant relationship with extension management ability of training organizers. Similar finding was observed by Patel (2001).

5.3.11 Staff Facility and Extension Management Ability

With respect to staff facility, the data presented in Table 31 indicate that the correlation coefficient was –0.120 indicating negative non-significant relationship with extension management ability of training organizers. It was striking to note that the negative relationship indicates that training organizers with less number of staff members in their KVK proved more efficient as manager of KVK. In the present study overall management ability was measured by considering personal ability of planning, organizing, directing, communicating, controlling, leading, supervising and decision making of training organizers. It is natural that when you have small staff to administer, it becomes easier for you to systematize, answer, communicate, converse, correspond, direct, guide, oversee and motivate each member of the organization to perform their work effectively. This might be the probable reason to have high level of extension management ability among those training organizers who had less staff members in their KVKs. However, calculated 'r' value indicates that, this inclination was not observed up to the level of significance.

Hence, null hypothesis (H_{01}) in case of staff facility was accepted and it can be inferred that there was non-significant relationship between staff facility and extension management ability of the training organizers.

Similar findings were reported by Patel *et al.* (1997) and Patel (2001).

5.3.12 Organizational Climate and Extension Management Ability

Table 31 shows that the correlation coefficient between organizational climate and extension management ability of training organizers was 0.524, indicating positive significant relationship between them. It shows that extension management ability of those training organizers was observed better, who had better and favourable organizational climate. The result indicates that training organizers with positive state of their employees' behaviours such as communication, problem solving, accountability, quality and performance, proved as favourable situation to perform as effective extension managers. In other words, training organizers who tried to create right environment for co-workers and simplicity by understating the true value of co-workers for the development of organization with the help of establishing proper communication, solving barriers and liability had high level of extension management ability. This might be because of simple reason that manager of any organizations can be successful if high quality work is performed by employees. Another reason for significant relationship may be that friendly, peaceful, cooperatively organizational climate might have led training organizers towards high level of extension management ability.

Hence, the null hypothesis, (H_{01}) was rejected and it can be concluded that organizational climate had significant relationship with extension management ability of the training organizers.

This finding is in line with the findings of Rani (1985), Padheria (1989) and Patel (2001).

5.3.13 Organizational Facility and Extension Management Ability

With respect to organizational facility, the data presented in Table 31 indicate that the correlation coefficient was 0.537 indicating significant relationship with extension management ability of training organizers. It means that training organizers with good organizational facilities were proved as better managers than those of with less organizational facility.

To accomplish mandates of KVK, training organizers have to do many activities like training, publication of farm literature, edification and organization of demonstrations, e-extension, field visits etc. To carry out all these activities, they need many infrastructural facilities like land, laboratory, vehicles, audio-visual aids, reading materials, furniture, stationary and other communication facilities like telephone, xerox machine, fax, computer and internet connections. Thus, it was obvious that training organizers with all the required organizational facilities could manage their duty effectively. This might be the reason for the above relationship.

Hence, the null hypothesis, (H_{01}) was rejected and it can be inferred that there was significant relationship between organizational facility and extension management ability of training organizers.

5.3.14 Interpersonal Communication and Extension Management Ability

The data in Table 31 stated that correlation coefficient of interpersonal communication with extension management ability of the training organizers was 0.401 indicating highly significant relationship at 0.01 level of significance. The result states that management ability of those training organizers was observed better who had better interpersonal communication quality.

The interpersonal communication was measured in terms of degree of contacts of training organizers with their heads, directors and officers of state agricultural university, state agricultural department and other officials. Thus, it is understood that training organizer with frequent contacts with other officers gets higher opportunity to discuss new problems, developmental issues, budge related matter, innovations and many other useful matters.

The persons working in any organization as a manager need to engage in interpersonal communication to gain knowledge about another individual. Social Penetration Theory says that we attempt to gain information about others so that we can interact with them more effectively. We can better predict how they think, feel and act if we know who they are. Training organizers through frequent contacts with others might have gained this ability to achieve high level of extension management ability.

Hence, the null hypothesis (H_{01}) was rejected and it can be concluded that interpersonal communication had significant relationship with extension management ability of training organizers.

5.3.15 *Personality and Extension Management Ability*

The result shown in Table 31 indicates that the personality of the training organizers was significantly correlated with their extension management ability. In other words it can be said that extension management ability of training organizers was better who had extrovert type of personality. It is well known fact that the person who has to work to inspire several people towards constructive track should have established personality, which is not subjugated by immature pleasure force. He should illustrate maturity to deal people by understanding norms and limitations of any society. The training organizers, working at KVKs with definite quality to work effectively with rural people and group along with motivation, self-confidence, friendly attitude and likeness to lead social gathering were seemed to be better in their extension management ability.

Extrovert individuality helps an individual to do extension work very successfully because of permutation of useful characters and positive approach to deal with external reality and environment rather than inner feelings. In addition to this extrovert individuals are highly sociable, active, assertive, energetic, enthusiastic, outgoing, impulsive and emotionally expressive, friendly and talkative. Such positivism might have played important role to have better extension management ability among those training organizers who had extrovert type of personality.

Hence, null hypothesis (H_{01}) in the case of personality of the training organizers was rejected and it was concluded that the personality of the training organizers was significantly correlated with their extension management ability.

5.3.16 Attitudes towards Extension Work and Extension Management Ability

It could be seen from Table 21 that attitude towards extension work had positive and significant relationship with extension management ability of training organizers.

In any organization, the attitudes of it's employee towards their work has a great bearing on its success or failure. Thus, the favourable attitude of training organizers towards extension work might have psychologically favoured to do better in their job work resulting in better managerial ability of extension work. This finding is supported by Rani (1985) and Patel (2001).

Thus, the null hypothesis (H_{01}) in case of attitude of training organizers towards extension work was rejected and concluded that there was positive significant relationship between attitude towards extension work and extension management ability of training organizers.

5.3.17 Workload Perceived and Extension Management Ability

The data regarding workload perceived in Table indicate that the correlation coefficient value was 0.023 indicating non-significant with extension management ability of training organizers. However, positive relationship indicates that extension management ability of those training organizers was better who had high or reasonable workload. In spite of this, the null hypothesis, (H_{01}) was accepted because such trend was not observed significant. Based on result it can be concluded that there was no any relationship between extension management ability of the training organizers and their perceived workload.

The probable reason for above findings might be that all-most training organizers have to perform similar kind of workload because mandates given by authority to all KVKs are more or less same.

5.3.18 Job Involvement and Extension Management Ability

In respect to job involvement, the data in Table indicate that the correlation coefficient was –0.047 indicating non-significant relationship with extension management ability. In other declaration, it can be said that job involvement did not play any role on the extension management ability of training organizers.

The probable reason for above findings might be same as explained in previous table that almost training organizers have to involve with various kind of activity to perform similarly because directives given by authority to all KVKs are relatively same.

Hence, the null hypothesis, (H_{01}) was accepted and it can be concluded that job involvement was non-significantly correlated with extension management ability of the training organizers.

Similar results were reported by Veerbhadraiah (1983), Reddy and Jayaramaiah (1988) and Patel (2001).

5.3.19 Job Stress and Extension Management Ability

The data in Table 31 with respect to job stress show that, job stress was indicating negative significant relationship with extension management ability of the training organizers.

The job stress has been worked out as feedback of training organizers with regards to their emotional and/or physical pressure for their job. A hectic situation occurs when equilibrium is not experiential between demand imposed upon them by the situation and their capacity to cope with such requirements. The person who is head of KVK should have physically and mentally powerful outlook to do work in any adverse situation. He should be sufficient strong, punctual, nervousness-free, matured, accommodating and talented to do any type of job successfully with whatsoever accessible staff he has. It is difficult to do such work. Thus it was seen that extension management ability of those training organizers was seen better who had less job stress. Understanding this fact, it is advisable that training organizers working at KVKs should be given specific training to deal with high job stress.

Hence, the null hypothesis, (H_{01}) was rejected and it can be inferred that job stress had negative significant relationship with extension management ability of training organizers.

This finding is in conformity with the findings of Talukdar (1984), Murthy and Somasundaram (1989) Patel *et al.* (1998) and Patel (2001).

5.4 The Functional Relationship between Extension Management Ability and Independent Variables

In the previous sub-section, the relationship between independent and dependent variables were ascertained by computing correlation coefficients (r). The correlation coefficient value gives only the strength and direction of relationship between the two characters or variables, but it does not reflect on predictive ability of independent variables to the dependent variable. Hence, in order to access the amount of contribution or say influence or predictive abilities of each independent variables to the dependent variable, the stepwise regression analysis was carried out with the help of computer.

The stepwise regression as stated by Efroymson's (1962) that it is one such method, which has been widely adopted in multiple regression analysis. It has got the added advantage that at each stage of analysis; every variable is subjected to an examination for its predictive value.

The multiple correlations co-efficient (R) represents the correlation between the dependent variable and set of independent variables fitted in multiple regression equation. The co-efficient of determination (R^2) gives amount of variation accounted in dependent variables, when all independent variables were taken together in the equation and was tested with 'F' test for its significance.

The partial regression co-efficient ($by_1.j$) represents the change in dependent variable (y) with a unit change in independent variable (X_1) keeping other variables constant and it was tested with Student's 't' test for its significance.

Table 32: Step-wise multiple regression analysis of extension management ability of Programme Coordinators

Sr. No.	Independent variable	*Multiple co-relation co-efficient (R)*	*Co-efficient of determin-ation (R2)*	*'F' values*	*Partial regression co-efficient (b)*	*'t' value*	*Standard partial regression co-efficient (SPRC)*	*Rank*
1	Attitude towards extension work	0.675	0.455 (45.50 %)	131.952**	1.700	11.487**	0.443	I
2	Personality of the training organizers	0.736	0.542 (54.20 %)	92.781**	0.383	5.447**	0.285	II
3	Organizational climate	0.746	0.557 (55.70 %)	65.376 **	0.236	2.320*	0.155	III

* Significant at 0.05 level of probability.

** Highly significant at 0.01 level of probability.

The various independent variables had their own unit of measurement, which did not permit a comparison of the partial regression co-efficient ($by_i.j$) values. To facilitate comparison, the partial ($by_i.j$) values were converted into standardized partial ($b'y_i.j$) values, which were free from the units of measurements.

The independent variables were than ranked on the basis of standard partial regression coefficient ($b'y_i.j$) values (ignoring sign) to find out their relative importance in predicting the dependent variable.

The content of the Table 32 revealed that the variables were introduced stepwise in succession depending upon the contribution of each of the independent variable in explaining the variation on the dependent variable.

From the Table 32, it can be observed that out of 19 independent variables only 3 were acquainting influence on the extension management ability of the training organizers. All the three variables together were contributing 55.70 per cent variation as indicated by 'R^2' value for the extension management ability.

It can be inferred that 45.50 per cent variation in extension management ability of training organizers of KVKs was contributed by attitude toward extension work. However, attitude towards extension work + personality of the training organizers accounted for 54.20 per cent, where as, attitude towards extension work + personality of the training organizers + organizational climate accounted 55.70 per cent variation in extension management ability of training organizers of KVKs. The R^2 values at each stage of step up regression were found to be significant at 0.01 level of probability.

The partial 'b' values of these three variables were converted into standard partial 'b' values, which were 0.443 for attitude towards extension work, 0.285 for personality of the training organizers and 0.155 for organizational climate. The 't' value or partial 'b' were significant in case of all the three independent variables.

According to the highest to lowest standard partial 'b' the rank order were given first attitude towards extension work, second to personality of the training organizers and third to organizational climate.

Hence, null hypothesis (H_{02}), in case of independent variables like attitude towards extension work, personality of training organizers and organizational climate was rejected and for remaining independent variables, it was accepted.

It can be inferred from the above results that independent variables such as attitude towards extension work, personality of the training organizers and organizational climate contributed 55.70 per cent variation in extension management ability of the training organizers.

The result shown in above Table indicates that the extrovert personality of the training organizers with their positive attitude towards extension activity and favourable organizational climate together played major role in making training organizers efficient manager of the training centres. It is natural that man with such positivism will always get chance to use his assertive and mature personality with positive approach to deal with internal and external reality and environment through positive inner feelings. It is also obvious that training organizers with favourable state of mind within themselves and positive organizational climate will get enough confidence to work as an efficient manager in any organizations. This might be the probable reason to receive major contribution accounted by the variables like attitude towards extension work, personality and organizational climate on the extension management ability of training organizers of KVKs.

The findings are suggestive of the fact that for increasing the extension management ability of the training organizers, proper training to build up positive feelings towards extension activity and extrovert personality should be organized for training organizers, at the same time he and his staff members should also be trained in such a way that conducive organizational climate can be maintained to get best results.

5.5 Constraints Faced by Training Organizers in Managing KVK and Suggestions Offered by Training Organizers to Overcome the Constraints

5.5.1 Constraints Faced by Training Organizers in Managing KVK

In the present study, some constrains faced by the training organizers in managing KVK were also studied. They were requested to express the constraints faced by them. These constraints were ranked according to their importance based on mean score and are presented in Table 33.

A glance of the Table 33 shows that no promotion channel has been arranged for the staff of KVK, many post of the subordinate staff lying vacant, intermittent flow of the funds from organization to KVK head for carrying out mandated activities, fund for the extension activities and FLDs is not released in time by the council, frequent transfer of the staff from KVK to other department are the major constraints faced by the training organizers in managing the KVK. Further they opined that most of the staff is engaged with organizational work or say other than KVK activities, inadequate infrastructure facilities such as building, fencing, tube well, vehicle etc., unnecessary involvement of organizational head in taking decision regarding planning, execution and implementations of the KVK activities, inadequate required educational qualification among the subordinate staff, local organizational rules do not allow the KVK staff to work freely according to the guidelines of council were the other constraints experienced and expressed by the training organizers to restrict their extension management ability.

Promotion encourages employees to view the organization as one offering them career growth. An incentive like promotion increases participation rate of employees and helps to create healthy completion to develop organization. It is needed to adhere healthy behaviours of employee. The purpose of the incentive is to encourage employees to adopt positive behaviours or maintain an existing positive behaviour. Everyone who achieves a goal or maintains a behaviour should receive something. Many organizations also provide incentives merely for active and healthy participation of employee in the development of organizations. Promotion helps in making organization more effective and enhances responsiveness of the employees. For this reason, if channel of promotion does exist in KVK setup, authority should take care of this constraint expressed by the training organizers.

The another important constrain faced by the training organizers, was issue of the vacant posts, it is natural that to perform all the activities of KVK effectively, sufficient number of subordinate staff is needed by the manager or any organizers, understanding this, urgent appointment of subordinate staff on the vacant posts was realized by the training organizers. The high performance team of human is one of the most important resources in development of any public dealing organizations. The organization like KVKs are involved in many activities, so to carry out work as per the mandates, training organizers need necessary active human resources. One should not expect best result from training organizers without enough and skillful human resources. In spite of positive attitude towards extension activity and mature personality, training organizers cannot do all kinds of work without skillful force of colleagues. The organizations that pay attention to the key factor of giving professional force of co-workers to the managers can create a positive and productive environment for their organizations, that will drive overall performance and long-term success.

The high performance teams with the quality of autonomous team-work attitude and self-directed work nature can perform their various roles effectively. Such kinds of co-workers are highly motivated and self-sufficient - drawing on their own resources and expertise. Rather than providing direction to high performance teams, managers act as consultants or resources to be utilized, as and when needed. Their autonomy that they have over their work activities enhances their ability to influence management's decisions on a variety of operational and strategic issues. High-performance teams of co-workers usually achieve a high level of motivation. Motivation, in this regard is translated into zeal, drive or enthusiasm. They are impelled, intrinsically to improve their performance. This high-level of motivation is achieved mainly because members of high-performance oriented co-workers feel that they are individually powerful, in control, highly responsible, eventually accountable, genuinely committed, highly dependable, highly empowered with the requisite degree of professional autonomy, explicitly trusted and highly respected.

Table 33:Constraints faced by training organizers in managing KVK

N = 160

Sr. No.	*Constraints*	*Mean score*	*Rank*
1	Many posts of the subordinate staff lying vacant.	1.79	II
2	Majority of the subordinate staff not fulfilling the required educational qualification.	1.31	IX
3	Frequent transfer of the staff from KVK to other department.	1.52	V
4	Fund for the extension activities / FLDs is not released in time by the council.	1.58	IV
5	Most of the staff remains engaged with local organizational work and less for the KVK activities.	1.41	VI
6	No promotion channel for the KVK staff.	1.93	I
7	Inadequate infrastructure facilities such as building, fencing, tube well, vehicle etc.	1.38	VII
8	Unnecessary involvement of organizational head in taking decision regarding planning, execution and implementations of the KVK activities.	1.35	VIII
9	Unfair interference of the local organizational head should be removed.	1.29	X
10	Intermittent flow of the funds from organization to KVK head for carrying out mandated activities.	1.59	III
11	Pressure from the local politician to fetch more benefits from KVK schemes to their own jurisdiction.	1.13	XI
12	Local organizational rules do not allow the KVK staff to work freely according to the guidelines of council.	1.31	IX

Realizing this, constraints like inadequate number of subordinate staff, lack of qualified staff and unnecessary engagement of staff in other than organizational activities were faced by the training organizers of KVKS.

All over India, KVKs are being run by SAUs, NGOs, Vidyapiths and ICAR research stations, it is therefore, training organizers have to depend on the concern authority to utilize funds of KVK because funds are not being given directly to them. Within this circumstance, untimely availability and improper utility of needed funds were understood as constraints by the training organizers. In addition to this, engagement of KVK staff to do less work for KVK and more for local organizations, intermittent flow of the funds from organization to KVK head for carrying out mandated activities, untimely release of fund for the extension activities/FLDs, pressure from the local politician to fetch more benefits from KVK schemes to their own jurisdiction and dependency on the administrative procedure of local organizational to work freely according to the guidelines of council were another constrains faced by the training organizers.

Hence, null hypothesis (H_{03}) was rejected and it can be concluded that training organizers faced some constraints in effective management of extension activities of KVKs.

5.5.2 Suggestions offered by Training Organizers to Overcome the Constraints in Managing KVK

In the present study, after studying some constraints faced by the training organizers in managing KVK, they were also requested to express their valuable suggestions based on their practical experiences. These suggestions were felt important to study to provide feed back to the concerned authority for the effective results of existing KVKs as well as to provide some guideline for those KVKs which are under considerations to establish allover India. The suggestions were collected in three continuums and their mean scores were worked out to understand its importance. The suggestions were also ranked according to their mean score and are presented in Table 34.

Table 34:Suggestions offered by training organizers to overcome the constraints in managing KVK

N = 160

Sr. No.	*Suggestions*	*Mean score*	*Rank*
1	Vacant posts especially technical including training organizers should be filled up immediately.	1.63	I
2	Disciplines as well as qualification should be considered while appointing technical staff.	1.52	II
3	Career advancement scheme for the KVK staff should be implemented and frequent transfer of the KVK technical staff should be avoided.	1.40	III
4	Demonstration unit / training laboratory as well as infrastructural premises should be well equipped.	1.36	IV
5	KVK grant should be directly released to the KVK account for timely operation. Additional grant should be provided to the KVK for carryout extension activities	1.32	V
6	Sufficient grant especially for extension activities should be provided and such grant should be released timely.	1.30	VI
7	Power and authority should be delegated to the training organizers adequately, considering their responsibility.	1.27	VII
8	Vehicle facility especially for carrying extension activities in village situation is highly essential.	1.24	VIII
9	The KVKs should be restricted to perform only extension tasks. Administration and extension work should be separated accordingly.	1.21	IX
10	Modern information and communication technology should be made available to all KVKs.	1.19	X
11	A strong coordination and linkage on permanent basis should be established between research institutions and KVKs.	1.17	XI

w

contd...

12	All the necessary infrastructural facility and educational facility for the staff should be made easily available.	1.14	XII
13	There should be a provision for work motivation and incentives for the staff of KVKs.	1.11	XIII
14	Frequent transfer of the staff should be restricted.	1.09	IXV
15	Mechanism of strong linkages and coordination with other extension agencies is necessary.	1.07	XV
16	An adequate personnel management policy should be adopted.	0.62	XVI
17	Standard and uniform evaluation system for extension work should be suggested by the ICAR to all the KVKs.	0.38	XVII

Table 34 reveals major suggestions given by training organizers in performing their work as extension mangers of KVKs. These suggestions are listed as below:

1. Vacant posts especially technical including training organizers should be filled up immediately. Without adequate technical staff, one cannot achieve the objectives of the KVK successfully.
2. Disciplines of subject as well as needed qualification should be maintained, while appointing technical staff. Without appropriate discipline/qualification/ specialization, quality of extension work may suffer a lot.
3. Career advancement scheme for the KVK staff should be implemented and frequent transfer of the KVK technical staff should be avoided.
4. Demonstration-unit/training-laboratory as well as infrastructural premises should be well equipped. Without the proper practical training facilities, one cannot do practice skillfully.
5. KVK grant should be directly released to the KVK for timely operation. Additional grant should be provided to the KVK to carryout extension activities e.g. Agricultural Fairs, Exhibition, celebration of different days like farmers' day, world food day etc. and maintenances of infrastructural facility.

6. Sufficient grant especially for extension activities should be provided and such grant should be released timely. Insufficient grant and delay in release of grant largely influence on planning and implementing the extension programme timely.
7. Power and authority should be delegated to the training organizers adequately, considering their responsibility. Decentralization of power and authority may be helpful to maintain the standard / quality or work and to carry out the responsibilities effectively.
8. Vehicle facility especially for carrying extension activities in village situation is an essential requirement, because extension work of KVKs is entirely rural based.
9. The KVKs should be restricted to perform only extension tasks, administration and extension work should be separated accordingly.
10. Modern information and communication technology such as computer, internet, fax, zerox, mobile phone, etc. should be made available to all KVKs, which are very essential for the development of extension management.
11. A strong coordination and linkage on permanent basis should be established between research institutions and KVKs with the help of the modern information / communication technology which will reduce the communication gap between them.
12. All the necessary infrastructural facility and educational facility for the staff should be made easily available which indirectly influences in a positive manner on extension management work of KVKs.
13. There should be a provision for work motivation and incentives for the staff of KVKs, which may result in effective and efficient management of extension work.
14. Frequent transfer of the staff is harmful. It adversely affects the planning and implementation of pre-decided extension programmes. Thus, model criteria for transfer of the staff should be designed by the ICAR especially for the KVKs working under SAU and ICAR.

15. Strong linkages and coordination with other extension agencies is necessary for effective management of extension programme.
16. An adequate personnel management policy should be adopted for better management of extension work of KVKs.
17. Standard and uniform evaluation system for extension work should be suggested by the ICAR to all the KVKs.

Table 34 reveals that major suggestions given by training organizers in performing their work as extension mangers of KVKs were related to appointment of staff, sufficient and timely major and additional grants, infrastructural facility, incentives, vehicle facility, career advancement, restriction on frequent transfer, modern information communication technology, linkage with other organizations and personnel management policy. To get best results through KVKs, their practical suggestions should be considered by the concerned authority. In addition to this, while establishing new KVKs allover India, training organizers of new KVKs should be provided all the facilities to run the activities of KVK as per the desired goals and mandates.

6. SUMMARY AND CONCLUSION

This chapter includes in a nutshell the description of summary, conclusion, problems and suggestions for future thrust.

6.1 Summary

The vigorous attempt was realized by the Education Commission (1964-66) to establish a specialized institution to provide vocational education in agriculture and allied fields at the pre and post matriculation levels to cater to the training needs of a large number of boys and girls coming from rural areas. The recommendation of Education Commission was thoroughly discussed during 1966-1972 associating the Ministry of Agriculture, Ministry of Education, Planning Commission, ICAR and other institutions. Finally, the ICAR mooted the idea of establishing Krishi Vigyan Kendra (Farm Science Centre) as innovative institutions for imparting vocational training to the farmers and field level extension functionaries. As a result of this, in 1973 a committee headed by Dr. Mohan Singh Mehta was constituted and based on its report in 1974, it was decided to have Krishi Vigyan

Kendras as Farm Science Centres for speedy transfer of technology to the farmer's fields. At present, about 504 KVKs working in India have been designed to have vocational training for the practicing farmers in service field to those who seek self-employment. The KVK as a grass root level institution has to meet this gap, which is so vital for technical literacy and rapid transfer of technology.

Krishi Vigyan Kendra is such an organization, which deals with particular social system within which a number of groups of people like; specialists, administrators, technicians, farmers, extension workers, labourers and other beneficiaries are working and involving. Each group has its own characteristics, attitudes, traditions, values and forms of behaviour. The people for whose benefits, the extension activities are designed may not be readily interested due to multiplicity of factors like; likeness, inhibitions, taboos, suspiciousness, reservations and prejudices. It is therefore, expected that manager of the KVK should have enough understanding of such social aspects to get best results out of KVKs. Thus, effectiveness of the KVK mainly depends on how it is efficiently and successfully managed by its heads or training organizer. The training organizers have to perform multifold roles of planner, organizer, director, communicator, controller, human relation maintainer, supervisor, coordinator, leader and decision-maker. It was, therefore, felt necessary to examine existing extension management abilities of training organizers of the KVKs. For this, present study on extension management abilities of training organizers of KVK was undertaken with following objectives.

1. To study personal, job related, organizational and socio-psychological characteristics of the training organizers of KVKs.
2. To measure the managerial ability of training organizers of KVKs.
3. To find out the relationship between selected independent variables and managerial ability of the training organizers of KVKs.
4. To estimate the functional relationship between managerial ability and various independent variables.
5. To study the constraints faced by training organizers of KVKs in the management of different activities of KVKs and seek their suggestions.

6.2 Review of Literature

On the basis of literature reviewed related to the study, a theoretical orientation was developed and various concepts were operationalized. Based on assumption, tentative conceptual model was developed and with the help of theoretical orientation, three statistical null hypotheses were also formulated in view of the above objectives.

6.3 Methodology

The present study was carried out in all eight zones of India, where Krishi Vigyan Kendras are in operation in transfer of technology process. The Ex-post facto research design was applied for the study. A random sample of total 160 training organizers of KVKs from the different zones of India was selected for this study. The methodological procedure consisted of dependent and independent variables. The independent variables studied were Age, Educational Qualification, Status of Health of training organizer, Type of Family, Size of Family, Native Place, Experience as a training organizer of KVKs, Total working experience in KVK, Total Training received, Span of Control, Availability of Technical Staff, Organizational climate, Organizational facilities, Inter-Personal communication, Personality of the Training Organizers, Attitude towards extension work, Work load, Job involvement and Job stress. These variables were measured by using well-developed scales developed by various social scientists with due modifications. The extension management ability of the training organizers of KVK was dependent variable, which was measured by using the scale developed by Patel (2001) with slight modifications. The data were collected by personal interview as well as mail questionnaire. The data thus, collected were classified, tabulated and analyzed in order to make the finding meaningful. The statistical measures, such as percentage, mean score, standard deviation, coefficient of correlation, stepwise regression and standard partial regression coefficient were used.

6.4 Hypotheses of the Study

H_{01}: There will not be any relationship between personal, job oriented, socio-psychological and organizational character of training organizers of KVKs and their extension management ability.

H_{02}: There will not be functional relationship between extension management ability of training organizers and various independent variables.

H_{03}: There will not be any constraints faced by training organizers of KVKs in managing KVK activities.

6.5 Major Findings

The following conclusions were drawn based on findings of the study :

6.5.1 Profile of the Programme Coordinators of KVK

1. Majority (58.75 per cent) of the training organizers had old age, followed by 31.88 per cent with middle age, while only 9.37 per cent of them had young age.
2. Majority (71.88 per cent) of the training organizers were educated up to doctorate level, followed by 19.37 per cent with master degree, while few of them (8.75 per cent) were educated up to bachelor degree only.
3. Two fifth (40.00 per cent) of the training organizers realized that their health was good, followed by 36.25 per cent and 21.25 per cent of the training organizers perceived that they were having average health and very good health, respectively. Equal percentages (1.25 per cent) of the training organizers felt that their health was poor or very poor.
4. Majority (80.00 per cent) of the training organizers had the nuclear type of family, while only 20.00 per cent of them had joint type of family.
5. Majority (60.00 per cent) of the training organizers had up to four members in their family, while 32.00 per cent of them had five to eight members in their family and only 8.00 per cent had above eight members in their family.
6. Majority (67.5 per cent) of the training organizers of KVKs had urban background, whereas 32.5 per cent of the training organizers of KVKs were from the rural area.
7. Majority (73.76 per cent) of the training organizers had medium level of total experience. The low and wide intensity of experience was observed among 14.37 per cent and 11.87 per cent of the training organizers, respectively.

8. More than half (51.88 per cent) of the training organizers had medium level of total experience in KVK on a different positions, whereas 25.00 per cent and 23.12 per cent of them had wide and low level experience, respectively.
9. Three-fourth (76.25 per cent) of the training organizers have received less training, whereas slightly less than one fourth (23.75 per cent) of them received high training.
10. Majority (80.62 per cent) of the training organizers had medium level of span of control, whereas 13.75 per cent and 5.63 per cent of them had low and high level of span of control, respectively.
11. Exactly half (50.00 per cent) of the training organizers opined that the strength of staff provided them was medium, whereas, 26.25 per cent of training organizers opined that existing staff strength was as per the requirement, whereas 23.75 per cent of them felt that insufficient staff was provided to their KVKs.
12. Majority (69.38 per cent) of the training organizers had perceived medium level of favourable organizational climate, while 16.25 per cent and 14.37 per cent of them had low and high level of favourable perception for their organizational climate.
13. Majority (74.38 per cent) of the training organizers felt that existing organizational facility with in their KVK was medium. On the other hand, low and high levels of organizational facility were perceived by 5.00 per cent and 20.62 per cent of the respondents, respectively.
14. Majority (58.13 per cent) of the training organizers had medium level interpersonal communication, whereas, 23.75 per cent of them had low level and only 18.12 per cent of them had high level of interpersonal communication.
15. Majority (76.88 per cent) of the training organizers exhibited extrovert personality.
16. Majority (78.75 per cent) of the training organizers of KVKs had favourable to highly favourable attitudes towards extension work.
17. Majority (60.00 per cent) of the training organizers had perceived desirable workload, whereas, 39.38 per cent of

them perceived heavy workload, only 00.62 per cent of them had perception of light workload.

18. Majority (80.00 per cent) of the training organizers were reasonably favourable, followed by 11.25 per cent with less and 8.75 per cent with high favourable towards their job.
19. Majority (73.12 per cent) of the training organizers experienced medium level of job stress, whereas 13.75 per cent with low level of job stress and 13.13 per cent of them had perception of high level of job stress.

6.5.2 Extension Management Ability of the Programme Coordinators

1. Slightly more than half (51.25 per cent) of the training organizers had high level of planning ability, followed by 33.75 per cent and 15.00 per cent with medium and low level of planning ability, respectively.
2. Majority (85.00 per cent) of the training organizers had medium to high level of organizing capacity in relation to extension management ability.
3. Majority of the (76.87 per cent) training organizers had high level of ability of directing their subordinates, followed by 13.75 and 9.38 per cent of them were with medium and low level of ability of directing their subordinates for efficient extension management activities.
4. Majority of the training organizer (58.13 per cent) had high level of ability of communicating, followed by 21.25 per cent of them with low and 20.62 per cent of them were with medium level of ability of communicating, respectively.
5. Slightly less than half (48.12 per cent) of the training organizers had high level of controlling power, followed by 28.12 per cent and 23.76 per cent of them were with low and medium level of controlling power, respectively.
6. Great majority (91.87 per cent) of the training organizers had medium to high level of capacity to maintain human relation, followed by only 8.13 per cent of them had low level of human relation maintaining capacity.
7. Majority (61.88 per cent) of the training organizers had high level of feeling of lead taking behaviour, followed by 20.00

per cent and 18.12 per cent of them were with medium and low level of feeling of lead taking behaviour, respectively.

8. Majority (71.26 per cent) of the training organizers indicated high level of supervising capacities, followed by 20.62 per cent and 8.12 per cent of them had medium and low level of supervising capacities, respectively.
9. Majority of the training organizers were found with high level of coordinating ability (62.50 per cent), followed by 26.87 per cent of them with low and only 10.63 per cent of them with medium level of ability of coordination in concern with personnel working in the KVKs.
10. Slightly less than half (45.00 per cent) of the training organizers had high level of decision making capacity, followed by 40.62 per cent of them with medium and only 14.38 per cent of them with low level of decision making capacities, respectively.
11. Majority (68.12 per cent) of the training organizers were observed with the medium level of overall extension management ability, while 18.75 per cent with high and 13.13 per cent with low level of overall extension management ability.

6.5.3 Relationship between the Characteristics of the Programme Coordinators and their Extension Management Ability

1. The personal variables like education, health status and native place of training organizers were observed positively, while age was observed negatively significant with their extension management ability. The variables viz., size of family and type of family were non-significant with extension management ability of training organizers.
2. The Job related variables like total experience as TO, total experience in KVK and training received were negatively non-significant with extension management ability of training organizers.
3. The organizational variables of training organizers like organizationalclimate,organizationalfacilityandinterpersonal communication were observed positively significant with their extension management ability. While, other organizational

variables like span of control was observed non-significant, only staff facility was observed negative but significant with their extension management ability.

4. The socio-psychological variables of training organizers like personality and attitudes towards extension work were observed positively significant, while job stress was observed negatively significant with their extension management ability. The other socio-psychological variables; workload and job involvement observed non-significant with extension management ability.

6.5.4 The Functional Relation between Extension Management Ability and Independent Variables of the Programme Coordinators

1. The 45.50 per cent variation in extension management ability of training organizers of KVKs was contributed by attitude toward extension work. However, attitude towards extension work + personality of the training organizers accounted for 54.20 per cent where as, attitude towards extension work + personality of the training organizers + organizational climate accounted 55.70 per cent variation in extension management ability of training organizers of KVKs.

6.5.5 The Constraints Faced by the Programme Coordinators

1. Lack of promotion channel for the staff of KVK, many post of the subordinate staff lying vacant, intermittent flow of the funds from organization to KVK head for carrying out mandated activities, pressure from the local politicians to fetch more benefits from KVK schemes to their own jurisdiction, funds for the extension activities and FLDs are not released in time by the council, frequent transfer of the staff from KVK to other department are the major constraints faced by the training organizers in managing the KVK.
2. Further they opined that most of the staff is engaged with organizational work or say other than KVK activities, lack of adequate infrastructure facilities such as building, fencing, tube well, vehicle, unnecessary involvement of organizational head in taking decision regarding planning, execution and

implementations of the KVK activities, inadequate required educational qualification among the subordinate staff, local organizational rules do not allow the KVK staff to work freely according to the guidelines of council are the other constraints experienced and expressed by the training organizers to restrict their extension management ability.

6.5.6 The Suggestions Expressed by the Programme Coordinators to Overcome Constraints

1. The major suggestions given by training organizers in performing their work effectively as extension mangers of KVKs were related to appointment of qualified technical staff, sufficient and timely availability of main and additional grants, infrastructural facility, incentives, vehicle facility, career advancement, restriction on frequent transfer, modern information communication technology, linkage with other organizations and personnel management policy. To get best results through KVKs, their practical suggestions should be considered by the concerned authority. In addition to this, while establishing new KVKs allover India, training organizers of new KVKs should be provided all the facilities to run the activities of KVK as per the desired goals and mandates.

6.6 Empirical Model

Base on results of relationship of independent variables with extension management ability of training organizers, an empirical model has been prepared.

It was seen that personal variables of training organizers of KVKs such as; young age, higher education, vigorous status of health and rural native place, while organizational variables like; conducive organizational climate, needed organizational facility and better interpersonal communication, whereas socio-psychological variables for instance; extrovert personality and positive attitudes towards extension work and low level of job stress played significant role on their extension management ability.

6.7 Implications

The present study was highlighted direction on the new area in which, research work was needed to carry out. The following implications emerged from this study:

1. A definite approach to craft India as developed nation in the shortest possible time is realized by a rapid agricultural growth. To accomplish the admired dream of father of nation; Mahatma Gandhi, we have to work with commitment in taking over newer information and technology to our millions of farmers who work hard vigorously in the sun to feed the nation. The world is shifting very rapidly. Technology of today becomes outdated tomorrow and agriculture is not exception to this. Agriculture plays an imperative role in Indian economy. Today in first line of transfer of technology mission, Krishi Vigyan Kendra needs to play a role of key extension system to develop the farming communities.
2. The findings of the study reveal that, to manage and run KVKs as per its mandate, there is need to appoint young, talented, well qualified, dedicated, enthusiastic training organizers to reinforce role of KVKs.
3. The results indicates that training organizers with rural native background were observed more active and commanding in managing extension activities and KVK, thus it is suggested that person with rural background or feelings to work in rural and semi urban area should be appointed to work as training organizers of KVKs. A Training organizer with rural background or feeling to work with farmers knows and understands better about farming situation prevailing in rural area.
4. During the study it was seen that training received by the training organizers did not exhibit any impact in improving their extension management ability. The training is said training if it shows any impaction the performance of any individual on his job. There is no meaning of any training, if it is not mental or physical skill improvement oriented. It is seen that many training organizations are failed to prove their ability in bringing desirable improvement on the performance of the trainees on their job. In such a situation,

in spite of exposure of person under many training one can not expect better results. This was observed in the present study. This may be due to the reason that the general training does not support much to management aspects. Thus, the training organizers with general training mostly on subject matters will not improve their skills of managerial ability of extension work. Thus, there is need to give special trainings which can prepare training organizers as good manager.

5. The positive relationship indicates that extension management ability of those training organizers was seen better who had more workers to look after. The training organizers with more members to control directly by them were more active in managing KVKs. It indicates that training organizer who had adopted strategy to maintain direct control and observation on all members of their KVKs had better extension management ability. Thus, it is recommended that training organizers should try to maintain direct contact with each member of KVK for effective management.
6. It was seen that extension management ability of those training organizers was observed better who had better and favourable organizational climate. The training organizers with positive state of their employees' behaviours such as communication, problem-solving, accountability, quality and performance got favourable situation to perform as effective extension managers. In other words training organizers who tried to create right environment for co-workers and simplicity by understating the true value of co-workers for the development of organization with the help of establishing proper communication, solving barriers and liability had high level of extension management ability. Thus, training organizers are recommended to maintain conducive organization climate to improve their level of extension management ability to manage KVKs in effective manner.
7. To accomplish mandates of KVK, training organizers have to perform many activities like training, publication of farm literature, edification and organization of demonstrations, e-extension, field visits etc. For this, they need many infrastructural facilities like land, laboratory, vehicles, audio-

visual aids, reading materials, furniture, stationary and other communication facilities like telephone, zerox machine, fax, computer and internet connections. Thus, it is recommended to provide all the facilities to each training organizer to manage their duty effectively.

8. The interpersonal communication is key component in development of any organization. It is degree of contacts of head of organization with other officials. It is needed to gain knowledge from another individual to predict how they think, feel and act. Such act provides ability to achieve high level of extension management ability. Thus, it is recommended to develop better interpersonal and interorganizational communication ability among the training organizers of KVKs.
9. The extension management ability of training organizers was observed better who had extrovert type of personality. The training organizers, working at KVKs with definite quality to work effectively with rural people and group along with motivation, self confidence, friendly attitude and likeness to lead social gathering, were seemed to be better in their extension management ability. It is thus, dreamed that each KVK of India should have such kind of training organizers. For this psychological training should be given to training organizers to develop extrovert individuality to make them able to deal with external reality and environment rather than inner feelings. In addition to this, such trainings help them to be highly sociable, active, assertive, energetic, enthusiastic outgoing, impulsive and emotionally expressive and open-minded.
10. The attitude towards extension work was seen positive and significant with extension management ability of training organizers. In any organization, the attitudes of it's employee towards their work has a great bearing on its success or failure. Thus, some strategy should be there to select and appoint person as training organizer of KVK with favourable attitude towards extension work.
11. The job stress was seen negative significant with extension management ability of the training organizers. The job

stress is feedback of training organizers with regards to their emotional and/or physical pressure for their job. A hectic situation occurs when equilibrium is not experiential between demand imposed upon them by the situation and their capacity to cope with such requirements. The person who is head of KVK should have physically and mentally powerful outlook to do work in any adverse situation. He should be sufficient strong, punctual, nervousness free, matured, accommodating and talented to do any type of job successfully with whatsoever accessible staff he has. It was seen that extension management ability of those training organizers was seen better who had less job stress. Understanding this fact it is advisable that training organizers working at KVKs should be given specific training to deal with high job stress.

12. The results indicates that variables such as attitude towards extension work, personality of the training organizers and organizational climate contributed 55.70 per cent variation in extension management ability of the training organizers, thus authority of KVK should try to do something considering these variables to create ideal and conducive situation in KVKs.
13. In the present study, some constrains faced by the training organizers in managing KVK were also studied. It was realized that no proper promotion channel has been arranged for the staff of KVK, many post of the subordinate staff are lying vacant, intermittent flow of the funds is not that effective from organization to KVK head for carrying out mandated activities, pressure from local politicians to fetch more benefits from KVK schemes to their own jurisdiction, fund for the extension activities and FLDs is not released in time by the council and frequent transfer of the staff from KVK to other department are the major constraints, which need to solve immediately.
14. During the study it was seen that there is perception of training organizers that most of the staff is engaged with organizational work (other than KVK activities), at the same time inadequate infrastructure facilities such as building,

fencing, tube well, vehicle etc. were also realized. Such problems need to resolve.

15. The KVKs are being managed by various organizations, so unnecessary involvement of organizational heads in taking decision regarding planning, execution and implementations of the KVK activities needs be resolved.
16. The another important constrain faced by the training organizers, was issue of the vacant posts, It is natural that to perform all the activities of KVK effectively, sufficient number of subordinate staff is needed by the manager of any organizers, understanding this, urgent appointment of highly qualified subordinate staff on the vacant posts should be realized. The high performance team of human is one of the most important resources in development of any public dealing organizations. The organization like KVKs are involved in many activities, so to carry out work as per the mandates, training organizers need necessary human resources. One should not expect best result from training organizers without enough and skillful human resources. Inspite of positive attitude towards extension activity and mature personality, training organizers cannot do all kinds of work without skillful force of colleagues. Organizations that pay attention to the key factor of giving professional force of co-workers to the managers can create a positive and productive environment for their organizations that will drive overall performance and long-term success.
17. All over India, KVKs are being run by SAUs, NGOs, Vidyapiths and ICAR research stations, it is therefore, training organizers have to depend on the concern authority to utilize funds of KVK because funds are not being given directly to them. Within this circumstance, untimely availability and improper utility of needed funds were understood by the training organizers. In addition to this, engagement of KVK staff to do less work for KVK and more for local organizations, intermittent flow of the funds from organization to KVK head for carrying out mandated activities, untimely release of fund for the extension activities / FLDs, pressure from the local politician to fetch more benefits from KVK schemes to

their own jurisdiction and dependency on the administrative procedure of local organizational to work freely according to the guidelines of council were another considerable issues to resolve.

18. There should be a provision for work motivation and incentives for the staff of KVKs.
19. The KVKs should be restricted to perform only extension tasks. Administration and extension work should be separated accordingly.
20. Modern information and communication technology such as Computer with all the software and accessories, internet, fax, zerox, mobile phone, printer, etc. should be made available to all the KVKs which are very essential for the modern development of extension management. It will be useful to establish strong coordination and linkage on permanent basis between research institutions and KVKs.
21. All over India, most of the KVKs are being established in rural or semi-urban areas, in this situation, dependent family members and children of staff working in KVKs face difficulty of good quality education, entertainment and some other facilities, authority should do something for this key issue and the necessary infrastructural facility and educational facility should be made easily available.

6.8 Suggestions for Future Thrust

In the light of findings of the study, the following studies can be undertaken to explore more in the area of internet exposure.

1. The present study was conducted in 160 KVKs of India, as this study is the first of its kind to know the extension management ability of training organizers. This study opens possibilities to carry out such studies on other training organizers of remaining KVKs of India.
2. The study was carried out under certain limitations of time and resources available with researcher, covering only training organizers of KVKs. However, such study can be carried out among other staff members of KVKs of India also.
3. This investigation was limited to only extension management ability of training organizers, but other aspects like impact of

internet exposure on the personality development of training organizers can also be included in new study.

4. Such study can be conducted on the extension educationists working in Agriculture, Horticulture, Animal Husbandry and other developmental departments of the Gujarat and India also.
5. To understand feelings of the farmers about the extension management ability of KVKs, one opinion study can also be conducted.
6. Latest agricultural knowledge and skill are the important factors to make training organizers as able and effective managers of KVKs, thus to understand existing level of knowledge and skill about scientific crop production, animal husbandry, dairy, value addition in agriculture, agricultural disaster management, sustainable agriculture, compost farming, integrated pest management, integrated nutrient management, integrated crop management, agricultural Government strategy, WTO, agricultural export, quality production, precise farming, hi-tech farming etc., a study can be conducted.
7. The Computer has become an integral part of daily life, thought it is not being used by many extension personnel. The use of computers and information technology can make training organizers of KVK effective and efficient managers, in spite of this, many training organizers of KVKs do not prefer to use computers and information communication technologies due to computer anxiety. Thus, a study to know level of computer anxiety among the training organizers can also be undertaken.
8. There are wide scope of internet for agricultural extension, thus a study on the awareness, attitude and views of the training organizers about internet can also be undertaken.
9. In the present study, some of the variables have been studied as independent variables. Some others independent variables which can play a role on the extension management ability of the training organizers, so new study can be undertaken by adding those untested and unstudied variables.

10. Some other aspects of training organizers such as extension productivity, role expectations, role performance, role-taking empathy etc. may be undertaken.
11. Such study on coordination, research-extension linkage and effectiveness or impact of KVK can be undertaken.
12. Such studies should be repeated after some lapse of time.

REFERENCES

1. Alvarez, A. and Arias, C. (2002). Diseconomics of size with fixed management ability. American J. of Agri. Econ., 85 (1) : 134-142.
2. Anonymous (1998). Faculty development in extension management. University News, 36 (44) : 27.
3. Anonymous (1998)*. Communication excellence in management, University News, 35 (50) : 51.
4. Anonymous (1999). Guidelines for the KVK Manager, New Delhi.
5. Anonymous (1999)*. Productivity. Agricultural Today, 2 (3): 58-59.
6. Anonymous (2000). Green revolution, the need of the hour. Agricultural News, 9 (2) : 65.
7. Anonymous (2000)*. Green revolution, the need of the hour. Agricultural News, 9 (2) : 68.
8. Anonymous (2001). News Digest, NDDB, Anand, Dec. 2001.
9. Anonymous (2002). Green revolution, the need of the hour. Agricultural News, 6 (3) : 63.
10. Anonymous (2005). Annual report 2004-05, Department of Agricultural Research and Education, Ministry of Agriculture, Govt. of India, ICAR, New Delhi.
11. Anonymous (2005)*. Speech of Mr. Sharad Pawar, Union Minister of Agriculture, Govt. of India, National Conference on KVKs – 2005, NASC Complex, New Delhi, October 27-28.
12. *Appley, S. A. (1990). Motive strength and leadership style as predictors of managerial effectiveness in selected community colleges. Dissertation Abst. International, 40 (9-A) : 4812.

13. Ardeshana, V. R. (1989). Teacher's attitude towards teaching at the College of Agriculture, Junagadh. M. Sc. (Agri.) Thesis (Unpublished), GAU, Sardar Krushinagar.

14. *Barnard, C. (1938). Social system theory. In : Reddy, P. N., Tripathi, P. C. and Appannaiah, H. R. 1990. Essential of management. Bombay, Himalaya Publishing House, pp. 24-25.

15. Bhandari, M. C. (1998). Professionalisation of cooperative management : Cooperative perspective. J. of Coop. Management, 5 : 15-17.

16. Bhatia, B. S.; Prem, K. and Chawla, A. S. (1990). Management of rural development (Ed.). New Delhi, Deep and Deep Publication, pp. 1-18.

17. Bhatia, B. S.; Prem, K. and Chawla, A. S. (1990)*. Management of rural development (Ed.). New Delhi, Deep and Deep Publication, pp. 1-18.

18. Bhimjiani, A. M. (1978). Study of personality characteristics as related to job performance of agricultural extension officers in Gujarat. M. Sc. (Agri.) Thesis (Unpub.), G. A. U., Navsari.

19. Bhople, R. S. and Palaspagar, V. W. (1996). Appraisal of managerial ability of farm supervisors, PKV. Res. J., 20 (2) : 161-162.

20. Borkar, M. M. and Chimurkar, B. S. (1996). Diagnostic ability of scientists engaged in T and V programme. Mah. J. of Ext. Edu., 15: 198-202.

20. *Brech, E. L. F. (1972). Administration and management in : Reddy, P. N., Tripathi, P. C. and Appannaiah, H. R., (1990). Essential of management. Bombay Himalaya Publishing House-9.

21. Chandel, R. S. (1974). A Handbook Agricultural Statistics, Achal Prakashan Mandir, Kanpur.

22. Chaudhary, P. K.; Sharma, S. S. and Gour, R. A. (1989). Job performance and job satisfaction of agricultural development officers. Mah. J. of. Ext. Edu., 8 : 245-247.

23. Chari, A. P. and Nandapurkar, G. G. (1989). A scale to measure managerial ability of farmers. Mah. J. of. Ext. Edu., 6: 163-198.

24. Chaya, P. (1974). An investigation into certain psychological characteristics of an effective schoolteacher, Ph. D. Thesis (Unpub.), Kanpur, University, Uttar Pradesh.

25. Chitra Pathak; Manish Kumar and Kwatra, J. K. (2005). Extension Training for Agricultural Development. Agri. Ext. Review, 17: 17-22.

26. Daftur, D. N. (1986). Delegation of authority scale. Department of Psychology, M. S. University, Baroda In : Pestonjee, D. M. 1989. Second Handbook of Psychological and Social Instruments (Ed.) New Delhi, Concept Publishing Company.

27. Dakhore, K. M. and Bhilegaonkar, M. G. (1988). Job satisfaction of veterinari extension personnel. Mah. J. of. Ext. Edu., 7 : 75-80.

28. Das, P. (2003). National Agricultural Technology Project (NATP). Indian Council of Agricultural Research. Proceedings of 26th meeting of the project management committee (PMC) of NATP, September, 2003.

29. Dasgupta, A (1969). Indian business management, Dept. of Business Management and Industrial Administration, University of Delhi : 60.

30. *Davis, R. C. (1951). Management defined In : Aggarwala, D. V. 1995. Principles practice techniques. New Delhi, Deep & Deep Publications, pp. 26.

31. *Drucker, Peter (1954). The practice of management. New York, Harper and Row Publishers.

32. *Drucker, Peter and Fayol, H (1954). Administration and Management. In : Reddy, P. N., Tripathi, P. C. and Appannaiah, H. R. 1990. Essentials of management. Bombay, Himalaya Publishing House, pp. 7-8.

33. *England, G. W.; Agrawal, N. C. and Trerise, R. E. (1971). Union leaders and Managers : a comparison of value system. Industrial Relation, 10: 211-214.

34. *Fayol Henry, (1949). General Industrial management. London SRI ISSAC Pitman and Sons.

35. Gardebroek, C. L. and Lansink, A. O. (2003). Estimating farm productivity differentials using panel data. J. of Agri. Econ., 54 (3): 397-415.

36. *Gantt, H. L. (1919). Early management approaches In : Reddy, P. N., Tripathi, P. C. and Appannaiah, H. R. 1990. Essential of management. Bombay Himalaya Publishing House, pp. 20.

37. *Gilbreth, F. (1915). Early management approaches In: Reddy, P. N., Tripathi, P. C. and Appannaiah, H. R. 1990. Essentials of management. Bombay, Himalaya Publishing House, pp. 20.

38. *Gilbreth, F. (1915)*. Early management approaches In : Reddy, P. N., Tripathi, P. C. and Appannaiah, H. R. 1990. Essentials of management. Bombay, Himalaya Publishing House, pp. 45.

39. *Gullick, L. (1925). Process management In : Reddy, P. N., Tripathi, P. C. and Appannaiah, H. R. 1990. Essential of management. Bombay. Himalaya Publishing House, pp. 21-22

40. *Hellriegel, D and Slocum, J. W. Jr. (1975). International Climate: Measures, research and contingencies. Academy of Management J., 17 : 225-280.

41. Hodge, B. J. and Johnson, H. J. (1970). Management defined In: Aggrawal, D. V. 1995. Management–Principles–Techniques New Delhi Deep & Deep Publications, pp. 25.

42. Igodan, Charis O. (1987). Job stress and burnout in extension work. Ind. J. of. Ext. Edu., 23 (1 & 2) : 1-6.

43. Islam, M. M. and Alli, M. H. (1991). Impact of training as perceived by the immediate workers. Ind. J. of Ext. Edu., 27 (1 & 2) : 81-87.

44. Jaggi, B. L. (1978). Management leadership styles in Indian work organizations. Indian Manager, 9 (2) : 139-156.

45. Jain, Uma R. (1981). Life styles of Indian managers : A study of district collectors. Ph. D. thesis (Unpub.), Indian Institute of Management, Ahmedabad.

46. Jondhale, S. G.; Ingle, L. A. and Pathak, U. N. (2000). Impact of Krishi Vigyan Kendra Training on adoption of improved practices of summer groundnut. Mah. J. of. Ext. Edu., 19 : 109-112.

47. Kalra, B. S. and Mishra A. K. (2005). Village level framework for watershed management. Agri. Ext. Review, 17 : 23-26.

48. Kareem, K. A. and Jayaramaiah, K. M. (1997). Problems perceived by the staff in the planning and implementation of integrated Rural Development Programme. Madras Agric. J., 84 (7) : 354-356.

49. Kerlinger, F. N. (1976). Foundation of Behavioural Research, New Delhi, Surjeet Publication, pp. 198-204.

50. Khare, H. C.; Jaiswal, D. K. and Mishra, P. K. (1987). Role performance of SMSs in T and V system. Maha. J. Ext. Edu., 6: 199-200.

51. Kholn, M. L. and Schooler, C. (1973). Occupational experience and psychological functioning. An assessment of reciprocal effects. American Sociological Review, 38 : 97-118.

52. Koontz, H. D. and O'Donnell, C. J. (1972). Principles of management An analysis of managerial functions Tokyo, McGram Hall, Kogakusha, pp. 5-20.

53. Koontz, H. D.; O'Dennell, C. J., and Weihrich, H. (1980). Management Auckland, McGraw Hill International Book Company, pp. 1-15.

54. Laharia, S. N. (1978). A study of personal and organizations variables influencing the productivity of agricultural scientists. Ph. D. (Agri.) Thesis (Unpub.), Haryana Agricultural University, Hissar.

55. Lee, Jung Hwan and Kim Song Soo (2004). The road ahead for Korean agriculture soft landing on a new plateau in the WTO. Seoul, Korean Rural Economic Institute.

56. Lodalh, T. M. and Kejner, M. (1965). The definition and measurement of job involvement. J. Appl. Psychology, 49: 24-33.

57. Malik, C. P. (1988). A study on job satisfaction of the teachers of Gujarat Agricultural University, Anand. M. Sc. (Agri.) Thesis (Unpub.), G. A.U., Sardar Krushinagar.

58. Malik C. P. and Patel, H. L. (1990). Factor related with job satisfaction in University. Mah. J. of. Ext. Edu., 9 : 42-45.

59. Mangala, Rai (2005). Annual Report of ICAR, Department of Agriculture (DARE), Ministry of Agriculture, GOI, New Delhi.

60. Mangala, Rai (2006). An inaugural speech at 76th Annual general meeting of the ICAR society, ICAR, New Delhi.

61. *Mayo, Elton. (1932). Human relation movement. In: Reddy, P. N., Tripathi, P. C. and Appannaiah, H. R. 1990. Essentials of management. Bombay, Himalaya Publishing House, pp. 22-23.

62. *McGreger, D. (1967). The professional manager. New York, McGraw Hill, pp. 5-7.

63. McGreger, D. (1971). The human side of enterprise. In: Rechards, M. D. and Neilander, W. A. (Eds.) Bombay, D. B. Taraporewala Sons & Co. Ltd.

64. Mee, J. F. (1963). Management defined In : Aggarwal, D. V. 1995. Management: Principles-Practices-techniques. New Delhi, Deep & Deep Publications, pp. 24.

65. Mehta, S. L. (1998). Convocation address in 23rd annual convocation at Gujarat Agricultural University, Sardar Krushinagar.

66. Monappa, A. and Saiyaddain, M. S. (1997). Personal management. New Delhi, McGraw Hill, pp. 327.

67. Morell, R. W. (1969). Management : Ends & Means. Sanfransisco, Chandler Pub. Co. pp. 134-143.

68. Murthy, R. and Somasundaram, D. (1989). Factors associated with the role performance of Village Extension Officers. Maha. J. Ext. Edu., 8: 28-34.

69. Nathall, P. L. (2001). Managerial ability a review of its basis and potential improvement using psychological concepts. J. of Agric. Econ., 24 (3) : 247-262.

70. Nikhade, D. M. and Kitey, P. V. (1984). Role performance of village level workers in agricultural development programme. Mah. J. of. Ext. Edu., 3 : 111-112.

71. *Niles, M. C. (1956). The Essence of Management Bombay, Orient Longman, pp. 20-24.

72. Padheria, M. M. (1989). Scientific productivity of academic staff from Marathwada Agricultural University, Parbhani : a system analysis. Ph. D. (Agri.) thesis, Marathwada Agricultural University, Maharashtra.

73. Padheria, M. M.; Soni, M. C. and Kulkarni, R. R. (1992). Prediction of scientific productivity of academicians. Guj. J. Ext. Edu., 1 : 1-4.

74. Pal, S. and Singh, Alka (1997). Agricultural Research and Extension in India: Institutional structure and investments. Policy Paper. No. 7, NCAR & PR. New Delhi. pp. 15-18 & 59-63.

75. Parmar, S. D. (2005). Publication behaviour of the Agricultural Scientists of Anand Agricultural University, Anand. M. Sc. (Agri.) Thesis (Unpub.), A.A.U., Anand.

76. Patel, C. S. (1990). Important qualities of agricultural college teacher as perceived by students and teachers of agricultural college. Guj. J. of Ext. Edu., 1 :93.

77. Patel, D. J. (1990). A study on teaching effectiveness of teachers working in Gujarat Agriculture University with special reference to faculty of agriculture. Ph. D. thesis (Unpub.), G.A.U., Anand.

78. Patel, D. P. (1992). Effectiveness of mass media on psychological attributes of agricultural teachers in Kheda district of Gujarat state. M. Sc. (Agri.) Thesis (Unpub.), G.A.U., Sardar Krushinagar.

79. Patel, A. A. (1993). Study on Research management ability of Research Scientist working as Head of Department/Station/Project/Scheme in G.A.U., Ph. D. Thesis (Unpub.), G.A.U., Anand

80. Patel, A. A.; Patel, M. R. and Desai, C. P. (1998)*. Research management ability of the heads of research departments in Gujarat Agricultural University. G AU. Res. J., 23 (1) : 56-59.

81. Patel, A. A.; Patel, M. R. and Desai, C. P. (1999). Job related and socio-psychological factors affecting research management ability of the heads in Gujarat Agricultural University. G AU. Res. J., 23 (2): 77-80.

82. Patel, G. J. (2001). A study on extension management efficiency of Training Organizer in KVK. Guj. J. of Ext. Edu., Dec.-2001. Vol. 14: 18.

83. Patel, G. J. (2001). A study on extension management ability of Training organizers on KVKs. Ph. D. Thesis (Unpub.), G.A.U., Sardar Krushinagar.

84. Patel, G. J. and Patel, A. A. (2003). Extension management power of TOs of KVKs in Gujarat state. Guj. J. of Ext. Edu., Dec. 2003. Vol. 18: 57-59.

85. Patel, G. J.; Soni, M. C. and Thakkar, K. A. (2004). A scale to measure extension management ability of training organizer of KVKs. Guj. J. Ext. Edu., 15 : 29-32.

86. Pawar, Sharad (2006). Presidential speech of Agricultural Minister at 76th Annual general meeting of the ICAR society at the NASC complex in New Delhi.

87. *Pearsons, T. (1954). Essays in Sociological theory Glencoe. The free press, pp. 275-286.

88. Popat, M. N. (1991). A study on productivity of SMSs working under T & V system in Gujarat State. Ph. D. Thesis (Unpub.), G.A.U., Anand.

89. Prasad, C. (1991). International conference on extension strategy for minimizing risk in rainfed agriculture. Ind. J. Ext. Edu., 27 (1 & 2): 1-12.

90. Rani G. Janshi (1985). Scientific productivity of agricultural scientists : An activity analysis approach. Ph. D. (Agri.) Thesis (Unpub.), Andhra Pradesh Agricultural University, Hyderabad.

91. Rao, R. and Mohan, R. (1986). A study on job satisfaction of teachers of Andhra Pradesh Agricultural University and factors associated with job satisfaction. M. Sc. (Agri.) Thesis (Unpub.), A.P.A.U., Hyderabad.

92. Rao, R. R. (1986). A study of job satisfaction of Andhra Pradesh Agricultural University and associated with job satisfaction. M. Sc. (Agri.) Thesis. (Unpub.), APAU, Hyderabad.

93. Reddy, M. S. and Jayaramaiah, K. M. (1988). Relationship between socio-psychological characteristics and the job effectiveness of VEWs. Ind. J. of. Ext. Edu., 24 (3 & 4) : 88-89.

94. Reddy, M. S. and Jayaramaiah, K. M. (1990). Relationship of selected variables on productivity of village extension officers. Ind. J. Ext. Edu., 26 (1 & 2) : 78-81.

95. Reddy, R. and Rao, B. V. (1998). Role performance and job satisfaction of village extension officers working with intensive agricultural extension programme (T & V system). In : Desai, G. R. and Reddy, M. R. (1986). Ed. Studies on T & V Extension in India, NIRD, Hyderabad, 2 : 80-81.

96. Reddy, M. N. and Sinha, B. P. (1990). Factors influencing the role conflict among the scientists in an agricultural university system. Mah. J. of. Ext. Edu., 9 (1 & 4) : 1-4.

97. *Rogers, E. M. (1962). Diffusion of innovation, New York, The free press of Glencoe, pp. 214-217.

98. Sah, A. K. (1999). Challenges before agricultural extension into 21st century. Agril. Ext. Review, 11 (5) : 1-8.

99. Sakaria, V. B. (1988). Role expectation, role preference, role performance and training needs of SMSs working under T & V system in Gujarat State. Ph. D. (Agri.) Thesis (Unpub.), G.A.U., Junagadh.

100. Samantha, D. K. (1985). Perception of work environment by agricultural research service scientists of ICAR institute. Ind. J. Ext. Edu., 21 (3 & 4): 53-61.

101. Samanta, D. K. (1991). Agricultural extension in changing world perspective. New Delhi, Uppal Publishing House, pp. 1-20 & 121-143.

102. Sawant, A. G. (2004). Role of informatics and bio-informatics in extension education in India. Asian J. of Ext. Edu., 23 (1) : 22-41.

103. Scanlan, B. K. (1973). Principles of management and organizational behaviors Canada, John Wiley & Sons Inc.

104. Sharma, M. L.; Sharma, P. N.; Jaiswal, P. K. and Sengar, R. S. (1988). Role expectation and role performance of rural agricultural extension officers of T & V system in M. P. Indian J. of. Ext. Edu., 24 (3 & 4): 75-78.

105. Shah, U. B. (2006). A study on level of Internet exposure of Teachers of Anand Agricultural University, Anand. Ph. D. Thesis (Unpub.), A.A.U., Anand, Gujarat.

106. Shivram, M. and Dalal, Y. C. (1994). Lekh Lakhvani Kala. (Art of writing : Article) Pravin Prakashan, Rajkot, Gujarat, p. 19.

107. Singh, J. P. (2000). WTO and India's agricultural export perspectives. Manage Ext. Res. Review, 1 (2) : 1-11.

108. Singh, R. (2001). Rural infrastructure, agricultural development and poverty in India. An inter state study. J. of Rural Development, 23 (1) : 31-57.

109. Singh, S. D.; Joshi, K. J. and Sharma, N. K. (2004). Mass media exposure among the horticultural development officers in Haryana. Rural India, 67 (5) : 90-94.

110. Singh, Baldeo and Padaria, R. N. (2004). Frontiers of scientific quests today in agricultural extension. Asian J. of Ext. Edu., 22 (1): 104-121.

111. Singhi, P. M.; Wadwalkar, S. and Kaur, G. (1982). Management of Agricultural Extension. Monogram No. 96; Indian Institute of Management, Ahmedabad, pp. 1-12.

112. Sendecor, G. W. and Cochran, W. G. (1967). Statistical methods VI. Ed. Oxford & IBI Publishing Co. Pvt. Ltd. New Delhi.

113. Soni, N. V. (1990). Content analysis of Krushi Go Vidya and Krushi Jivan farm magazines: Special problem. M. Sc. (Agri.) Thesis, G.A.U., Junagadh.

114. Srivastava, V. L. (1981). Contours of management. New Delhi Classical Publishers, pp. 1-8.

115. Stagner, R. P. (1956). Perception : applied aspects. In the psychology of industrial conflict. John. Wiley & Sons, New York.

116. Sudesh, Rani (2005). A study on gain in knowledge and attitude of trained women towards fruit and vegetables preservation training of Vadodra district. M. Sc. (Agri.) Thesis (Unpub.), A.A.U., Anand.

117. Sukhadia, A. G. (2005). Author's motivation towards writing articles in Krushi Jivan magazine. M. Sc. (Agri.) Thesis (Unpub.), A.A.U., Anand.

118. Sulaiman, V. R. and Hall, A. (2004). Towards extension – plus: Opportunities and Research, Policy Brief 17 New Delhi.

119. Swanson, B. E. (2004). Extension strategies for poverty alleviation in a global economy. Paper presented at the international seminar on "Best practices in poverty alleviation through education for the new century". Suwom, South Korea.

120. Talukdar, R. K. (1984). Productivity of agricultural development officers in Haryana : A factor analysis study. Ph. D. Thesis (Unpub.), H.A.U., Hissar.

121. *Taylor, F. W. (1915). Early management approaches In : Reddy, P. N., Tripathi, P. C. and Appannaiah, H. R. 1990. Essentials of management. Bombay, Himalaya Publishing House, pp. 19-20.

122. Terry, G. R. (1968). Meaning of Management. In: Reddy, P. N., Tripathi, P. C. and Appannaiah, H. R. 1990. Essentials of management. Bombay, Himalaya Publishing House, pp. 20.

123. Thakkar, D. (1986). A study on the adoption of recommended summer groundnut technology by the farmers of Anand talukas of Kheda district of Gujarat state. M. Sc. (Agri.) Thesis (Unpub.), G.A.U., Anand.

124. Tolmay, V. L. and Hernandez, P. (2001). Resistance to biotic and abiotic stress in the Triticeae, proceedings of fourth international Triticeae Symposium, Cordoba, 135 (2&3) : 239-242.

125. Trip, G.; Thijssen, G. L.; Renkeema, J. A. and Huirne, R. B. M. (2003). Measuring managerial efficiency : the case of green house growers J. Agric. Econ., 27 (2) : 175-181.

126. Tripathi, B. D. (2000). Agricultural Extension personnel in new millennium : A prospective view. Manage Ext. Res. Review, 1 (2): 12-23.

127. Unger, P. W.; Skidmore, E. L. and Carter, M. R. (1995). Conservation tillage in the southern United States Great Plains, conservation tillage in temperate agro-ecosystems, U. S. A. : 329-356.

128. Urwick, L. and Gullick, L (1925). Process Management In : Reddy, P. N., Tripathi, P. C and Appannaiah, H. R. 1990. Essential of management, Bombay, Himalaya Publishing House, pp. 21-22.

129. Van de Ban, A. W. (2004). Changes in society require changes in agricultural extension : some observations, Asian J. of Ext. Edu., 23 (1): 1-17.

130. Van den Ban, A. W. and Hawkins, H. S. (1996). Agricultural extension. London, Blackwell Science Ltd., pp. 1-6 & 233-266.

131. Vance, S. (1956). Meaning of Management. In : Reddy, P. N., Tripathi, P. C. and Appannaiah, H. R. 1990. Essentials of management. Bombay, Himalaya Publishing House, p. 3.

132. Veerbhadraiah, V. (1983). Time management, job involvement and job performance of extension supervisors. Thesis Abstract, 9 (2) : 124.

133. Veerbhadraiah, V. and Jalihal, K. A. (1983). A study on time utilization of deputy directors and assistant director of agriculture on different activities. Mysore J. Agric. Sci., 9 (2) : 342-347.

134. Venkateshwarlu, N. (1988). Foundation course of agricultural research project management course. No. 28, Lecture Notes, NAARM, Hyderabad.

135. Verma, O. S. (1997). Management of Extension system : Authority, responsibility and accountability. J. of Ext. System, 13 (1 & 2) : 20-35.

136. Vinayangam, S. and Samanta, R. K. (2004). Assessment of work environment among trainers in KVKs. Asian J. of Ext. Edu., 23 (1): 129-134.

137. Vroom, V. H. (1964). Work and motivation New York, John Wiley & Sons Inc.

138. Wase, R. B. (2001). Knowledge and adoption of farmers about Jayanti Chilli cultivation M. Sc. (Agri.) Thesis (Unpub.), P.D.K.V., Akola, Maharashtra.

139. Woodworth, R. S. (1979). Assessment of personality (Scale): Personality. New Delhi, Oxford.

140. Wu-Sx; Deva Doss, S.; S and Lu, Y. C. (2003). Estimation and decomposition of tech. efficiency for sugar beet farms. Applied Economics, U.S.A., 35 (4) : 471-484.

** Original not seen*

APPENDICES

Appendix- I

INTERVIEW SCHEDULE EXTENSION MANAGEMENT ABILITY OF THE TRAINING ORGANIZERS OF KVK

Name of the respondents: ______________________________

Name of KVK : ______________________________

Whether KVK belongs to SAU / ICAR / NGO / Deemed University

PART - I

INDEPENDENT VARIABLES

(A) Personal information

1) Age of the respondents in completed years: ______________
2) Educational qualification: ______________________
3) What do you feel about your health (Very good/Good/Average/ Poor/Very poor)
4) Type of family (a) Joint (b) Nuclear
5) Total no. of family member ______________________
6) Native place (a) Rural (b) Urban

(B) Job related information

1) Total experience as training organizers ____________ years
2) Total working experience in different cadres (KVK) __________
3) Total No. of training received during service period

   Short term training (up to 7 days) Nos. _______

   Medium term training (8 to 15 days) Nos. _______

   Long term training (above 15 days) Nos. _______

   Total Nos. _______ years

PART -II
ORGANIZATIONAL INFORMATION

(A) Span of control

Give total no. of staff working under your guidance

Cadre	*No.*
Technical	
Administrative	
Auxiliary	
Total	

(B) Staff facility: what do you feel about staff facility provided to you in different cadres to carryout extension activities of KVK (√) your response

sr. No.	*Category*	*More than required (excess) (2)*	*As per requirement (sufficient) (3)*	*Less than required (Insufficient) (1)*
1	Technical			
2	Administrative			
3	Auxiliary			

(C) Organizational Climate: Please indicate your response by putting (√) against each statement which explains the level of organizational climate of your KVK.

Sr. No.	*Statements*	*Response*			
		Highly (4)	*Quite (3)*	*Some what (2)*	*Not at all (1)*
1	Staff is generally interested in welfare and happiness of each other				
2	Staff is clear about the goal and objectives of KVK				

contd...

3	Differences and disagreement are not avoided but sorted out				
4	Staff is satisfied here with their job				
5	Staff feels responsible for effective functioning of KVK				
6	Staff has confidence and trust in each other				
7	Information is shared by the staff freely				
8	Staff is generally gives ideas and suggestions to each other				
9	There is an atmosphere of warmth and supporting in the KVK				
10	Staff is generally bothered if some one faces difficulty				
11	There are opportunity to use and develop capacities of staff				
12	Staff takes initiative to improve the work of KVK				
13	There is a good prevailing team spirit				
14	Good work is always appreciated				
15	Staff is concerned with effective performance and use innovative was rather than referring the rules				

(D) Organizational facilities :

Please indicate which of the following facilities available with your KVK.

Sr. No.	*Facilities*	*Available*	*Easily available from other*	*Available with some difficulty*	*Not available*
A	Training facilities				
1	Audio-visual				
2	Training hall / lab.				
3	Training material				
4	Reading material				
5	Photographic facility				
6	Videography				
7	Hostel for trainers				
8	A-V equipment				
9	Others, if any				
B	Other facilities				
1	Office building				
2	Office furniture				
3	Telephone				
4	Fax				
5	Xerox				
6	Computer				
7	Internet connection				
8	Cyclostyle machine				
9	Typewriter				
10	Vehicle for extension work				
11	Others, if any				

(E) Interpersonal communication

How frequently do you discuss about your activities / problems with the following persons? Tick mark (√)) in the appropriate column.

Sr. No.	*Persons*	*Response*				
		Very frequently (5)	*frequently (4)*	*Some-times (3)*	*Seldom (2)*	*Never (1)*
1	Head of your organization					
2	Officers of ICAR					
3	Officers of SAU					
4	Officers of state Dept. of agriculture					
5	Officers of line department at district level					
6	Technical / research staff					
7	Administrative staff					
8	Others, if any					

PART–III

SOCIO-PSYCHOLOGICAL VARIABLES

1) PERSONALITY OF TRAINING ORGANIZERS

The following questions are relating to your personality. Please answer each statement by marking (√).

Sr. No.	*Questions*	*Answer*	
		yes	*No*
1	Do you like to talk before a group of people		
2	Are you some what suspicious of other people motives		
3	Do you always try to make others agree with you?		
4	Do you suffer from the feeling of inferiority?		
5	Do you make friends easily?		
6	Are you feeling easily hurts?		
7	Are you at home among strangers?		
8	Do you worry over what people think of you?		
9	Do you like to take the lead in a social gathering?		

2) Attitude towards extension work :

You are requested to check (√) appropriate response category against each statement expressing your attitude towards extension work.

Sr. No.	*Persons*	*Attitude towards extension work*				
		sA	*A*	*u*	*dA*	*sd*
1	Honestly speaking, I hate extension					
2	Extension work is the need of the hour.					
3	Money spent on extension work is sheer wastage.					
4	I really enjoy meeting farmers and doing extension work					
5	I am trapped in extension profession					

contd...

6	The future of the nation lies only in extension work					
7	Extension work is a thankless job.					
8	Extension work is a good medium of serving humanity					
9	Extension workers are considered as second class citizen.					
10	I have poor status in the eyes of relatives only because I am working as training organizer.					

SA = Strongly agree, A = Agree, U = Undecided, DA = Disagree, SD= Strongly disagree

3) Perceived of workload

What do you perceive about your workload? Please (√) against the options given below.

Light workload (1) ( )

Sufficient workload (2) ( )

Heavy workload (3) ( )

4) Job involvement

The following statements regarding orientation to job are given, please indicate your response about your job involvement by putting (√) against each statement.

Sr. No	*Statements*	*Response categories*				
		sA	*A*	*u*	*dA*	*sd*
1	I have other activities more important than my work.					
2	To me work is only a small part of my life					
3	Quite often I feel like staying at home instead of going for work.					
4	I used to care more about my work but now other things are important to me.					

contd...

5	We can measure a person pretty well by how a good job he does.					
6	I shall stay overtime to finish a job, even if I am not paid for it.					
7	I live, eat and breath for my job.					
8	Sometimes I would like to kick my self for the mistakes I make in my work					
9	For me nothing at work really go off quickly					
10	I used to be more ambitious about my work than I am now.					
11	I usually go for work a little early to get the things ready.					
12	I am really a perfectionist about my work.					
13	I avoid taking extra duties and responsibilities in my work.					
14	Most things in life are more important than work.					
15	I feel depressed when I fail at something connected with my work.					
16	Sometimes I keep myself awake at night thinking ahead to the next days work					
17	The major satisfaction in my life comes from my work					
18	I would keep working even if I do not get the money.					
19	I am very much involved personally in my work.					
20	The most important things that happen to me involves my work.					

SA = Strongly agree, A = Agree, U = Undecided, DA = Disagree, SD= Strongly disagree

5) Job stress

Given below are some situations, which arise in your life. These situations process problems, which disturb the balances of life and threaten the person resulting in loss of face, prestige or security in the job life. You are requested to (√) appropriate response category against each statement expressing your agreement or disagreement with statement.

Sr. No.	*Statements*	*Response Categories*				
		sA	*A*	*u*	*dA*	*sd*
1	My assignments are of monotonous nature					
2	I get less salary in comparisons to the quantum of my work.					
3	My decision and instructions concerning work among staff are properly followed.					
4	I do my work under tense circumstances.					
5	Owing to excessive workload, I have to manage with insufficient staff resources.					
6	I get ample opportunity to utilize my abilities and experiences independently.					
7	I have to do some work unwillingly owing to certain group / political pressure.					
8	Some of my colleagues and subordinates become happy on my unsuccessful events.					
9	My college does cooperate with me voluntary in solving work problems.					
10	I have to dispose my work hurriedly due to excessive workload.					
11	I am unable to perform my duties smoothly due to short of authorities.					
12	Working conditions are satisfactory level for welfare and convenience of staff.					

contd...

13	I often feel that this job has made my life burdens one.					
14	Being to busy with official work, I am not able to devote sufficient time to my domestic and personal problem.					
15	The mutual cooperation and team spirit among the staff are existing here.					
16	I have to do such work as out to be done by others.					
17	I am unable to carry out the work to my satisfaction due to excessive load of work and lack of time.					

PART – IV

1) Constraints faced by the training organizers :

Being a head of the KVK, you may experience Management / Technical / Administrative constraints which you find obstacles in the effective functioning of the Kendra. Please tick against your (√) response.

Sr. No.	*Constraints*	*Category*		
		Mf (3)	*f (2)*	*NA (1)*
1	Many posts of the subordinate staff lying vacant			
2	Majority of the subordinate staff not fulfilling the required educational qualification			
3	Frequent transfer of the staff from KVK to other department.			
4	Funds for the extension activities / FLDs not released in time by the council.			
5	Most of the staff is engaged with organizational work (other than KVK activities)			
6	No promotion channel for the KVK staff.			
7	Inadequate infrastructure facilities such as building, fencing, borewell, vehicle etc.			

contd...

8	Unnecessary involvement of organizational head in taking decision regarding planning, execution and implementations of the KVK activities.			
9	Unfair contact of the organizational head with subordinate staff.			
10	Intermittent flow of the funds from organization to KVK head for carrying out mandated activities.			
11	Pressure from the local politician to fetch more benefits from KVK schemes to their own jurisdiction.			
12	Organizational rules not allow the KVK staff to work freely according to the guidelines of council.			
13	If any, (you can add)			

MF = Most frequently, F = Frequently, NA = Not at all

1)	Please give your suggestions to overcome the constraints.
1.	
2.	
3.	
4.	
5.	

PART - V

Information pertaining to extension Management ability of training organizers

You have to perform many functions related to management of extension activities. Following are the statements regarding management functions. Please indicate your perception against each statement putting (√) marks.

1) Planning: How often do you plan about different functions?

Sr. No.	*Sub indicators with weightages*	*Always (3)*	*Some times (2)*	*Never (1)*
1	To determine the objectives			
2	To ascertain training needs			
3	To plan for transfer of technologies			
4	To plan for utilization of funds			
5	To plan for extension facilities			
6	To prepare annual plan of work			
7	To plan for monitoring work			
8	To plan for evaluation work			

(2) Organizing: How often do you organize different functions?

Sr. No.	*Sub indicators with weightages*	*Always (3)*	*Some times (2)*	*Never (1)*
1	To call regular meeting			
2	To provide work opportunity			
3	To provide work according to specializations			
4	To invite suggestions from subordinates			
5	To divide the activity into groups			
6	To delegate authority according to responsibility			

(3) Directing: How often do you direct subordinates about different functions ?

Sr. No.	*Sub indicators with weightages*	*Always (3)*	*Some times (2)*	*Never (1)*
1	To give clear instruction of work			
2	To give due credit to a person			
3	To appreciate good work done			
4	To give continuous information			

(4) Communicating: How often do you involve to communicate different functions ?

Sr. No.	*Sub indicators with weightages*	*Always (3)*	*Some times (2)*	*Never (1)*
1	To give timely instructions			
2	To listen to subordinates			
3	To select proper medium to instructions			
4	To receive feed back			

(5) Controlling: How often do you control different functions ?

Sr. No.	*Sub indicators with weightages*	*Always (3)*	*Some times (2)*	*Never (1)*
1	To monitor the work			
2	To visit accurately and properly			
3	To determine standard for work			
4	To evaluate the work			
5	To measure performance of staff			

(6) Human Relation: How often do you make personal touch with persons to carry out functions ?

Sr. No.	*Sub indicators with weightages*	*Always (3)*	*Some times (2)*	*Never (1)*
1	To make personal contacts			
2	To deal fairly			
3	To take interest in development of juniors			
4	To adopt participative approach			
5	To understand others behaviour			
6	To provide co-operation			

(7) Leading: How often do you take lead in different functions ?

Sr. No.	*Sub indicators with weightages*	*Always (3)*	*Some times (2)*	*Never (1)*
1	To promote team work			
2	To take initiative task			
3	To apply technical competency			
4	To have a sense of responsibility			
5	To inspire subordinates			

(8) Supervising: How often do you supervise different functions ?

Sr. No.	*Sub indicators with weightages*	*Always (3)*	*Some times (2)*	*Never (1)*
1	To provide wide counseling			
2	To focus on quality of work			
3	To give suggestions			

(9) Co-ordinating : How often do you co-ordinate different functions ?

Sr. No.	*Sub indicators with weightages*	*Always (3)*	*Some times (2)*	*Never (1)*
1	To establish linkages			
2	To integrate the extension work			
3	To establish mutual trust			
4	To get co-operation from outside			

(10) Decision making : How often do you take proper decision about different functions ?

Sr. No.	*Sub indicators with weightages*	*Always (3)*	*Some times (2)*	*Never (1)*
1	To take decision at proper time			
2	To find out facts for decision			
3	To adopt participative decision			
4	To implement decision			
5	To take decision free from biases			

Appendix - II

Instruments used for measurement of variables

The variables studied and instruments used for their measurement are summarized as follows :

Sr. No.	*Variables*	*Instrument used for measurement*
(A)	Dependent variables	
1	Extension management ability	Used scale developed by Patel (2001)
(B)	Independent Variables	
1	Age	Structured schedule on chronological age-scoring

contd...

2	Educational qualification	Structured schedule on academic degree-scoring
3	Health	Scale used as Popat (1991)
4	Type of family	Structured schedule on type of family-scoring
5	Size of family	Structured schedule on family members scoring
6	Native place	Structured schedule on native place-scoring
7	Experience as a head of KVK	Structured schedule on exp. as head-scoring
8	Total experience in KVK	Structured schedule on total exp. in KVK-scoring
9	Training received	Structured schedule in total training received-scoring
10	Span of control	Structured schedule on span of control total staff-scoring
11	Availability of technical staff	Structured schedule on technical staff available-scoring
12	Organizational climate	Jain Uma (1981)
13	Organizational facilities	Structured schedule on organizational facilities-scoring
14	Inter personal communication	Structured questionnaire on inter personal communication-scoring
15	Personality of Training Organizer	Scale developed by Woodworth (1979)
16	Attitude towards extension work	Scale developed by Daftur (1986)
17	Perceived work	Structured schedule on work load-scoring
18	Job involvement	Scale developed by Lodalh and Kejner (1965)
19	Job stress	Structured questionnaire on job stress-scoring

4 Construction of Scale to Measure Computer Anxiety/Nervousness

ABSTRACT

Emerging technologies are continually advancing the computer's capacity and usefulness making "the computer" a difficult term to define in the broadest sense; a computer is an information-processing machine. It can store data in form of numbers, letters, pictures or symbols and manipulate at great speed by following instructions that have been stores in the machine as programmes.

As we continue to barrel through the information age, it is hard to imagine conducting any research, educational, extension or business activity without computers. Each day millions of people working in offices and houses around the world depend on computer technology to do their jobs efficiently and economically. To truly understand the computer's history involves daunting journey through mathematic, physics and electrical engineering through binary code, Boolean logic, real time, magnetic core memories, floating-point numerical notation, transistors, semi transistors, conductors, integrated circuits and much more.

Luckily most office workers do not need to understand this complex history to use computers and the dizzying array to software programme they execute, when computers were first developed nearly fifty years ago. The people who programmed them considered the task quite maddening. Fortunately, learning to use a personal computer today is often as simple as spending a few hours reading an instruction manual or following hands on tutorial.

In recent years, computer technology has been incorporated into a wide range of consumers and industrial products. Computers are routinely used in world processing, E-mail, entertainment and offer applications that require repetitive tasks that can be automated. The present computer is a result of series of modification. The recent computer is super computer.

Earlier big area was need for computer co-accommodate in the offices, but due to the advancement made in computer, it has became smaller, faster and extremely capable information handlers. In 1976 a college dropout named Steve W. and a Teenagar named Steve Jobs founded affordable computers to easy use, named the personal computer. Nowadays, thousands of software are available to perform almost any imaginable task using built-in rules and procedures, these programme offer fast and efficient ways to conduct any business teaching, research, learning/education. Starting from registration for admission to be successful businessman, serviceman or for getting another remarkable achievement the computer is became a part and parcel of the life. In the field of education also it has became a part of the life of the academicians and students. It is used extensively by those students who are involved in research work. Especially agriculture education is concerned with the knowledge of diversified filed for which affection with computer is very much essential for the agricultural students. It is observed that due to computer anxiety, many students are not in position to take advantages of computer application for their development. It is therefore there is a need to examine the major problems and issues associated with students in working with computer.

The challenges in adopting computerization in day-to-day work are a prominent need of the time. To know existing sensitivity of the students towards computer in using this smart machine for the development of their carrier and education, reliable information on the anxiety or say nervousness of the students towards computer, systematically developed reliable scale was realized to develop. This can, therefore, be used to know their affiliation with computer to face the challenges in executing computer applications.

With a view to constructing a scale to measure the computer anxiety/ nervousness, present study was under taken with following specific objectives;

i. To construct and standardize a scale to measure computer anxiety/ nervousness of an individuals.
ii. To find out level of computer anxiety/nervousness of students studying in the agricultural college.
iii. To study the profile of the students and its relationship with their level of computer anxiety/nervousness.

The technique chosen to construct the attitude scale was the scale product method which is the combination of Thurston's and Likert's Techniques. In developing the scale 24 statements about computer anxiety/nervousness were collected. The data from 50 judges were arranged from which 16 statement were selected based on the median and Q values to constitute attitude scale.

By spilt half reliability test interval consistency of a scale was measured which came to 0.716. This study was conducted on students of the B. A. College of Agriculture, Anand Agricultural University, Anand-388110, Gujarat state during August, 2005. The results of students computer nervousness says that majority (81.00 per cent) of the students had medium to high degree of computer anxiety/nervousness.

The results regarding profile of the students indicated that 68.00 per cent of the agricultural students had below 23 year of age, slightly more than half of them (52.00 per cent) had first class with distinction in their academic performance, more than half (57 per cent) of their father's education was up to higher secondary and college level. About 54.00 per cent of the student had urban and 46.00 percent had rural background, majority (68.00 per cent) of them had nuclear type of family, aspiration to migrate aboard either for study or for permanent settlement (77.00 per cent), wish to get higher degree (78.00 per cent). Majority of the student did not possess computer at home (60.00 per cent), 60.00 per cent of them had e-mail ID and 29.00 per cent of then had everyday exposure of library.

The developed test to measure anxiety/nervousness is such that it indicates less anxiety/nervousness among persons with high score and more anxiety/ nervousness among persons with low score. Thus the result indicates that the computer nervousness was observed more among those students who had young age means studying in first or second years. The other variables such as academic performance, father's education, native place, type of family, wish to migrate abroad, wish to get higher degree, possession of computer at home, possession of E-mail/ID and library exposure did not have significant relationship with the computer anxiety/nervousness of the students.

1. INTRODUCTION

Computer is a machine that performs tasks, such as calculations or electronic communication, under the control of a set of instructions called a program. Programs within the computer are retrieved and processed by the computer's electronics. The program results are stored or routed to output devices, such as video display monitors or printers. Computers perform a wide variety of activities reliably, accurately, and quickly.

Computers are used for different purposes. In business, computers track inventories with bar codes and scanners, check the credit status of customers, and transfer funds electronically. At homes, tiny computers embedded in the electronic circuitry of most appliances control the indoor temperature, operate home security systems, tell the time, and turn videocassette recorders (VCRs) on and off. Computers in automobiles regulate the flow of fuel, thereby increasing gas mileage. Computers also entertain, creating digitized sound on stereo systems or computer-animated features from a digitally encoded laser disc.

Computer programs or applications exist to aid every level of education, from programs that teach simple addition or sentence construction to programs that teach advanced calculus. Educators use computers to track grades and communicate with students. With computer-controlled projection units, they can also add graphics, sound and animation to make communication effective. Computers are used extensively in scientific research to solve mathematical problems, investigate complicated data, or model systems that are too costly or impractical to build, such as testing the air flow around the next generation of aircraft. The military employs computers in sophisticated communications to encode and unscramble messages, and to keep track of personnel and supplies. In recent times, computers are at the base in e-governance, e-business, e-learning, e-mailing and such other cyber-based activities, thus revolutionizing the official as well as social dealings of the people.

2. THE FUTURE OF COMPUTER

In 1965, semiconductor pioneer Gordon Moore predicted that the number of transistors contained on a computer chip would double every year. This is now known as Moore's Law, and it has proven to be somewhat accurate. The number of transistors and the computational speed of microprocessors currently double approximately every 18 months. Components continue to shrink in size and are becoming faster, cheaper, and more versatile.

Computers will become more advanced and they will also become easier to use. Improved speech recognition will make the operation of a computer easier. Virtual reality, the technology of interacting with a computer using all of the human senses, will also contribute to better human and computer interfaces. Standards for virtual-reality program languages for example, Virtual Reality Modeling Language (VRML) are currently in use or are being developed for the World Wide Web.

Other exotic models of computation are being developed, including biological computing that uses living organisms, molecular computing that uses molecules with particular properties, and computing that uses deoxyribonucleic acid (DNA), the basic unit of heredity, to store data and carry out operations. These are examples of possible future computational platforms that, so far, are limited in abilities or are mostly theoretical. Scientists investigate them because of the physical limitations of miniaturizing circuits embedded in silicon. There are also limitations related to heat generated by even the tiniest of transistors.

Intriguing breakthroughs occurred in the area of quantum computing in the late 1990s. Quantum computers under development use components of a chloroform molecule (a combination of chlorine and hydrogen atoms) and a variation of a medical procedure called magnetic resonance imaging (MRI) to compute at a molecular level. Scientists use a branch of physics called quantum mechanics, which describes the behaviour of subatomic particles (particles that make up atoms), as the basis for quantum computing. Quantum computers may one day be thousands to millions of times faster than current computers, because they take advantages of the laws that govern the behaviour of subatomic particles. These laws allow quantum computers to examine all possible answers to a query simultaneously. Future uses of quantum computers could include code breaking and large database queries. Theorists of chemistry, computer science, mathematics and physics are now working to determine the possibilities and limitations of quantum computing.

Communications between computer users and networks will benefit from new technologies such as broadband communication systems that can carry significantly more data faster or more conveniently to and from the vast interconnected databases that continue to grow in number and type. With their increasing power and versatility, computers simplify day-to-day life. Unfortunately, as computer use becomes more widespread, so do the opportunities for misuse. Computer hackers-people who illegally gain access to computer systems-often violate privacy and can tamper with or destroy records. Programs called viruses or worms can replicate and spread from computer to computer, erasing information or causing malfunctions. Other individuals have used computers to electronically embezzle funds and alter credit histories. New ethical issues also have arisen, such as how to regulate material on the Internet and the World Wide Web. Long-standing issues, such as privacy and freedom of expression, are being reexamined in light of the digital revolution. Individuals, companies, and governments are working to solve these problems through informed conversation, compromise, better computer security and regulatory legislation.

3. COMPUTER ANXIETY/NERVOUSNESS

As a result of tremendous usefulness of computer to make life easier and meaningful, computer has become an important part of day-to-day life of many people, at the same time it is also seen that due to some fear and lack of confidence and awareness many people do not want to make use of computer in their life. Such behaviour in a broad sense is known as computer anxiety or nervousness.

Computer anxiety is a common emotional response to computers characterized by the fear that many people exhibit. Fear and anxiety toward subject matter are "conditions that tend to support negative learner attitudes and repel person's interest".

Anxiety usually occurs when something new is being learned. This causes resistance to change and has negative effects on cognitive performance. Individual learns to use computers often fear the unpredictability of computers, public exposure of ignorance, and threat of failure. These fears contribute to negative learner attitudes and are detrimental to learning. Feelings of anxiety toward computers and computer use, is common, affecting big mass of the population. Yet, with technology moving into libraries and classroom, and computers becoming an essential teaching tool, technophobes are struggling.

Anxiety by definition is intense dread, apprehension, or nagging worry. Computer anxiety is fear of impending interaction with a computer that is disproportionate to the actual threat presented by the computer. Computer anxiety is a concept specific anxiety type; that regularly occurs in a specific type of situation. Those who are computer anxious may experience fear of the unknown, feeling of frustration, possible embarrassment, failure and disappointment.

The computers have modernized the processing and transmission of information. Although the Agricultural Extension System is slow in adopting computer technology for its extension educational programs. In India the computer technology can be an invaluable asset to our educational programs raises a question: Are Extension agents anxious about the advent and use of computers? If so, they may not accept and use computers in their work. Computer-anxious people are those who fear using computers or become afraid at the prospect of using them. They do not choose to learn computer at all, they may have extreme negative feelings, similar to phobia. The computer-anxious people will avoid interaction with computers. In a study of computer attitudes, it was found that three-fourths of home economics professionals weren't comfortable around computers, and suggested that Extension employees must be allowed time and financial support to become computer literate.

Keeping in view an increasing importance of Information and Communication Technology (ICT) in agriculture, the newly emerging fields like Information Technology, Biotechnology, and Bioinformatics become very important day-by-day. The establishment of an agripolyclinics and agriinformatic centres is a need of the time. The knowledge of computer is an essential for students, especially for agriculture students. Information Technology becomes an essential part of Research, Teaching and Extension in the field of agriculture. In a golden era of Information Super Highway the basic knowledge of computer is an essential for all. Such knowledge is needed among any individuals to have more affiliation with computer and reduce their computer anxiety or nervousness. The person without knowledge of computer will be designated as "out dated" in days to come.

4. EXPLANATION OF PROBLEM

Computer mediated communication is fast becoming an intricate part of our society. Exposure to this new medium gives one the opportunity to acquire unlimited amounts of knowledge and a chance to communicate with others around the world. It is a fast way to create, send and consume new information. Computer mediated communication extends our mental capabilities and enhances our intellect. Yet there are many people who will be left behind in the technological quest for the faster and more efficient mode of communication.

Our access to people, places and information are changing. Libraries now use computer programs to catalog. A real change can be found in education. Despite income, school budgets or demographics soon all students will have access to information through the internet. E-mail is taking the place of inter-office correspondence. Business is rapidly becoming computerized. Students and workers will need to be comfortable with computers. As the academic and business environments continue to move forward in computer technology, the gap is widening for those people who experience computer anxiety.

Feelings of nervousness toward computers and computer use, is widespread, affecting personality development for those who want to be an important part of new millennium. Many college students experience some type of techno-phobia. It is genuine aversion toward computers and computer use. Does potential users' attitude affect or influence their computer usage? Will computer increase computer experience reduce anxiety? What can be done to alleviate computer anxiety? To answer these questions one must first understand the existing status of the computer anxiety.

The newly emerged concepts of virtual (class room) teaching, Distance Learning, On-demand Education and On-line Examination are completely based on computer and computer-aided technologies. For that the basic knowledge of computer is important at the same time it is also helpful to break down the computer fear from the student is of prime importance.

Implementation and adoption of computer can help not only to reduce paperwork in offices but it also helps an individual to make his work faster, easier and more meaningful. To accelerate rate of use of computer, positivism of an individual in terms of skill, knowledge and attitude is very much essential. To understand real gap among all this motivational factors of skill, knowledge and attitude, systematically well-developed scales or tests are needed. Looking to this fact, a scale was thought necessary to develop to measure computer Anxiety/ Nervousness of an individual with following objectives.

5. OBJECTIVES OF THE STUDY

The present study was undertaken with following objectives:

1. To construct and standardize a scale to measure to measure computer Anxiety/ Nervousness of an individual.
2. To find out level of computer Anxiety/Nervousness of students studying in the agricultural college.
3. To study the profile of the students and its relationship with their computer Anxiety/Nervousness.

6. REVIEW OF LITERATURE

6.1. Scale

Edward (1957) has discussed the problem of assessing attitude by direct questioning and by observation, and concluded that both the methods would give crucial classification of the attitude. He further added that for a quick and convenient as well as reliable quantitative measurement of attitude of large group, an attitude scale yield fruitful results. An attitude scale, therefore serves as a mean of assessing degree of positive or negative affect that individual is associated with some psychological objects. Among the technique available for the construction of scales, the oldest Thurston's Equal Appearing Interval Scale (1928) and the Likert's Summated Rating Scale (1932) are quite well known during present time also. The combination of Thurston and Likert's techniques, known as "Scale Product Method" is also used to construct an attitude scale.

Many attitude scales have been developed so far; among these some have used the "Scale Product Method" which combines Thurston's techniques of Equal Appearing Interval for selection of items and Likert's Summated Rating for ascertaining the response on the scale. Mathew and Reddy (1989), Popat (1990), Patel (2000) are among those who used the combination of Thurston and Likert's techniques, known as "Scale Product Method" to construct an attitude scale.

Patel (2000) concluded that attitude of beneficiary farmers towards National Watershed Development Project was positively and significantly associated with their extent of adoption of watershed management technology. The higher education, participation in social organizations, active involvement in various extension programmes, progressiveness, scientific orientation of the farmers may have jointly contributed in forming favourable attitude towards watershed development programme.

Patel (1991) found that majority (75.71 per cent) of the respondents had favourable attitude, whereas nearly equal number (12.39 and 11.90 per cent) of the respondents had high and less favourable attitude towards modern technology based agricultural programme of watershed development. Further he observed that farmers' attitude towards modern technology based programme was positively and significantly associated with adoption of new agricultural technology.

Patel *et al.* (1995) reported that farmers' attitude was found to be positively and significantly associated with their extent of adoption of new agricultural technology.

Karkar (1998) revealed that the level of attitude of farmers was highly and significantly associated with extent of adoption of modern rainfed agro technology.

6.2 Computer Anxiety

Pam Dupin-Bryant (2002) reported that the increased demand for adult computer training produces many issues and challenges for Extension educators. Computer anxiety is a psychological characteristic of adult learners that computer instructors will continually confront.

Hakkinen (1994) suggested that it is important to reduce anxiety because it has negative effects on learning. Computer anxiety can be significantly reduced if Extension educators create a comfortable learning environment in which students can have positive experiences with computers.

Tseng *et al.* (1997) reported that feelings of anxiety toward computers and computer use, is common, affecting 30 to 40% of the population.

Brown and Vician (1997) reported that computer anxiety has been associated with decreased use and worse, avoidance of information technology. Avoidance of computer use can seriously affect some students' academic progress which may cause lower performance in business settings and ultimately affect career opportunities.

Necessary and Parish (1989) shown a positive relationship exists between experience levels with computers and favourable attitude toward computers.

A pilot study done by Harris (University of Texas at Austin) and Grangennett (University of Nebraska at Omaha) made reference to several well known researchers. "Computer anxiety levels have been found to be better predictors of success in using computers that is extent of prior computer experience (Marcoulides, 1988), but computer anxiety scores are not related to amounts of computer experience (Rosen, Sears & Weil, 1987; Marcoulides). Computer experience appears to effect attitudes about computers, rather than computer anxiety (Gressard and Loyd, 1986, Igbaria & Chakrabarti, 1990).

DeLoughry (1993) reported that encourage students to explore, "People need the opportunity to play before they gain confidence." He quoted the words of Weil that schools, colleges and businesses allow students and employees to experiment with new computers and new software before they are incorporated into daily activities. This would allow the user the time to explore, "play", and feel comfortable.

Wlodkowski (1993) Computer anxiety is a common emotional response to computers characterized by the fear that many adults exhibit. Fear and anxiety toward subject matter are "conditions that tend to support negative learner attitudes and repel adult interest".

Hakkinen (1994) suggests that this interaction may incite a variety of emotional responses, including anxiety. The fear of computers interferes with the communicative nature of human-computer interaction.

Anxiety usually occurs when something new is being learned. This causes resistance to change and has negative effects on cognitive performance. Adults learning to use computers often fear the unpredictability of computers, public exposure of ignorance, and threat of failure. These fears contribute to negative learner attitudes and are detrimental to learning (Wlodkowski, 1993).

Anonymous (2004) More and more, professional success depends upon reasonable application of computer technology. However, not all people find easy access. Certainly, two major hindrances are computer anxiety and lack of computer self-efficacy.

Patel (2004) observed that majority of the postgraduate students having computer exposure had everyday exposure of Internet (53.33 per cent) and possession of at least one or more e-mail ID (81.66 per cent). Exactly half of them (50.00 per cent) had more than three years of exposure of Internet with everyday to at least once in a month contact of chatting facility of Internet. More than one third (35.00 per cent) were using e-mail facility at list twice in a week to receive or send mails. Major purposes expressed by Internet use postgraduate students were to collect research related information and to send or receive e-mails, while major activity carried our by them was writing useful information on separate papers. Major sources of Internet utilized by them were library or private cyber café/ Google and yahoo were the main search engines utilized by users to connect themselves with useful sites. Independents factors like age, education, exposure in extra curricular activities and library exposure were observed significant with Internet exposure of post graduate students.

Patel (2004) further more reported that majority of the computer exposed and internet user postgraduate students of Agriculture College were studying in M.Sc (Agri) (65.00 per cent). Majority of them had above 23 years of age (51.67 per cent), first class (51.67 per cent), above S.S.C level of education of their father (83.33 per cent), up to S.S.C or more than that level education of their mother (70.00 per cent), native place in rural area (51.66 per cent), nuclear and small size of family (51.66 per cent), aspiration to go abroad either for further study or for permanent settling (66.67 per cent), wish to get higher academic degree (51.67 per cent), low level of exposure of extra curricular activity and everyday exposure of library (75.00 per cent). The level of Internet exposure of the postgraduate students was positively and significantly related with their age, education, exposure in extra curricular activities and exposure of Library. The level of Internet exposure was observed better among those postgraduate students who were studying in Ph.D, having more age and higher level of involvement in extra curricular activities and library.

6.3 IT in Agriculture

Anonymous (2001) according to a media Matrix survey conducted it was seen that nearly 25 per cent of professionals who used IM (Chatting) at home have started using it to contact clients and colleagues.

Anonymous (2002) Assessment Bulletin volume-6 Number 1 November 2002 in this issue reported on the results of a survey of UMASS Amherst Freshmen and sophomores about their exposure. The survey force E-mail is commonly used for communication between professor and students, course websites and required internet research are also fairly common students tend to be exposed to course based interactive technologies less frequently and student interest in taking online courses is mixed few have actually taken such a course.

Anonymous (2002a) reported that information technology has been one of the most aspired fields in today's world integrating IT with agriculture will help any country to regulate its overall economy and trade. The different IT technologies like expert system in decision support system. Remote sensing, Gyandoot etc. have brought revolution in Indian agriculture Even corporate like ITC MSSL, HLL, are now looking forward to extract huge benefits out of this collaboration of IT with agriculture.

Anonymous (2002b) reported that seven in ten students used the campus library website for atleast some of their assignment and one-in five used it for most assignments college student find out about the library's website from multiple sources, including from their professors and teaching assistants (49%), by looking it up themselves (45 %) from classes about using the library (34%) and from libraries (27%). Among the students who did not use the campus library's website some (20%) did not know the library had a site, and some (29%) said it didn't have what they need, but less than half (43%) left other sites have better information. During their most recent electronic visit, most used full taxes of journal articles (67%) databases ad journal indexes (51%) and electrum books (21%). Few college students used any "Ask-a-librarian" services. Nearly 9 out of 10 students (89%) also use the campus library's print resources including books, journals, articles and encyclopedias. In addition to using the library's print resources they made photocopies from print resources and even print copies of electronic resources.

Anonymous (2003a) to succeed in a web-based course, students should be motivated and self directed the minimum required items to complete the course work at a distance successfully are (i) Seriousness: Online classes aren't for goof-offs who seek easy credits. Virtual students should expect to spend at least as much five on homework as those in traditional courses. (ii) Self disciplined: It is up to students to budget there tie and keep up with assignments. They must crease and stick to their own schedules (iii) Self reliance: The ability to undependably solve problems or research information is needed questions can be answered by e-mail but that takes time (iv) Careful reading skills : Because classroom lectures are replaced primarily by written work, students need to be careful and slow thoughtful readers, and (v) Computer skills: Students must be comfortable using computers, and the internet that includes E-mail web browsing downloading and word processing.

Anonymous (2003b) reported that in answer of the question to the teachers "When your student was deciding whereto go to college the most helpful information she/he received was from" out of 125 respondents, 14 said the internet, 9 said printed publications, 85 said campus visits, 3 said college admissions office, 3 said high school counselors/teachers and 11 said friends.

Barbara (1996) concluded that the more students are exposed to what the internet has to offer, the better and more interested users they become.

Curtis (2002) reported that the age of internet user respondents was quite varied ranges from 35 to 50 years the most prevalent online skills learned were computer application and software skills (64.00 per cent) as well as technical skills (50.00 per cent)

David (2000) reported that Edmonton public schools strives to ensure that students were provided with a safe and secure learning environment when engaged in educational assignment importing internet activities students use computers for activities that 90 hand in hand with our understanding of what constitutes a traditional childhood communicate and form relationship as children always have.

Drs Douglas W. Green and Thomas O'Brien (2002) reported that at first glance, the activity associated with the internet project had a little time giving direction and student were very active. Students were very eager to help each other and teachers spent most of their time facilitating student work students had many opportunities to tell teachers what they had found and it was common to hear teachers respond with comments such as "I didn't know that" But most of the assignments offered students some degree of choice, increasing their level of interest and providing the opportunity to relate to their experiences.

Hamizatul (1997), Yeap (1998) and Pang (1997) identified several factors that learn internet skills is even stronger because of "a shortage of teaching staff" as well as equipment in institutions of higher learning. He further observed that students face several major barriers to their successful use of library resources through internet were genuine, students perceive that access was denied because of the inability to access databases remotely due to password requirements and/or license restrictions. Difficulty searching and mitigating within the library and its websites costs of copying and printing at the library shortage of knowledge able librarians lack of the customer orientation they have come to expect as consumers

In Yeap's (1998) comparative study of the Arts and Science students at that the lack of exposure and shortage of equipments influenced to a great extent of internet Yeap found that the science students were more likely to use internet, because of their knowledge in internet application compared with the Arts students.

Jessic Donn *et al.* (2001) reported several possible explanations for graduate students greater acceptance of forming relationship on the internet Graduate students being older and therefore closer to the age of which any people marry, may have empathy for the desire to meet people. A second possible explanation of their findings was that graduate students may already be some what accustomed to using the internet for other purposes such as a making professional contacts and doing research.

Krishnatray and Kulshrestha (2002) concluded that among the internet uses college going boys and girls 6 per cent were boys and 40 per cent were girls. Most of them were graduates and three per cent were post graduates from the age group of 24 years more than 70 per cent of them had experience of three years, the parents of most students are graduates in the general streams and are engaged in government / private service or business.

Neil Marriott, Neil Selwyn and Prumarriot (1999) reported that it was encouraging that the vast majority of students in this study were found to be making some meaningful use of ICT. International students were found to use the university's ICT provision more than home students.

Teresa *et al.* (2003) indicated that when question asked students during their survey to specify their level of comport when seeking information from a variety of sources, including search engines a library web page a friend a professor/ TA/ GSA, a faculty or class website or the library Students responded that they were most comfortable seeking information from a search engine.

Yee (1998) studied the potential of e-commerce among teaching staff and student of a university of Sarawak the study found that users were not convinced with the needful and security of payment through internet.

Yeun *et al.* (1997) reported that total of 18 males and 19 females participated in the study. Compute ownership high comprised 89 per cent whereas internet access at home amounts to 35 per cent of the population.

Anonymous (2004) reported that internet based technologies are changing logistic from being a packaging and moving function in to an information business Electronic business integrates carriers with shippers via electronic ordering inventory decisions and product flow. Retailers and third party logistics firms are able to access their respective information system through the internet.

7. RESEARCH METHODOLOGY

7.1 Research Methodology to Develop Attitude Scale

This part of the assignment covers details about the data used in the study, manner and method through which the data were collected and it gives details of the method and procedure followed in construction of scale.

In this study an attempt has been made to develop a scale, which can scientifically measure person's computer Anxiety/Nervousness. Among the technique available for the construction of scales, the Thurston's Equal Appearing Interval Scale (1928) and the Likert's Summated Rating Scale (1932) are quite well known. Both the methods suffer from the limitations, the first one in getting discriminating response and second one in selection of items. Thus, technique chosen to construct the scale was of "Scale Product Method" which combines the Thurston's technique of Equal Appearing Interval Scale for selection of the items and Likert's techniques of summated rating for ascertaining the response on the scale as proposed by Eysenck and Crown (1949).

8. STEPS IN CONSTRUCTION OF ATTITUDE SCALE

8.1. Item Collection

The items making up scale are known as statements. A statement may be defined as any thing that is said about a psychological object. As a first step in the developing the scale, 24 statements about computer anxiety/nervousness were collected from the relevant literature, experts of computer cell and extension personals. The statements, thus selected were edited on the basis of the criteria suggested by Thurstone and Cave (1927), Wang (1932), Likert (1932) and Edward and Kilpatrick (1948).

8.2 Judge's Rating of Attitude Statements

In order to judge the degree of "Unfavorableness" to "Favorableness" of each statement on the five point equal appearing interval continuum a panel of 50 judges was selected. The judges selected for the study comprised extension educationist, computer specialists and statisticians with considerable practical experience in computer used from the Anand Agricultural University. The judges were visited personally along with letter of instructions to guide them for rating the statements in desired manner for each set of the statements as seen in Appendix-I.

8.3 Determination of Scale and Quartile Value

The five points of the rating scale were assigned score ranging from 1 for most unfavourable and 5 for most favourable. The based on judgment, the median value of the distribution, and the Q value for the statement concerned was calculated with the help of following formula.

$$S = L + \frac{0.50 - Pb}{Pw \times i}$$

Where,

S = The medium or scale value of the statement

L = Lower limit of the interval in which the median falls

Pb = The sum of the proportion below the interval in which the median falls

Pw = The proportion within the interval in which the median falls

i = The width of the interval and is assumed to be equal to 1.0 (one).

The inter-quartile range ($Q = Q_3 - Q_1$) for each statement was also worked out for determination of ambiguity involve in the statement.

8.4 Research Methodology to Employ Developed Scale

9. LOCATION OF RESEARCH AREA

After development of the scale to measure computer Anxiety/ Nervousness, it was implemented on small sample of the students of agricultural college. It was employed on the randomly selected 50 undergraduate students studying in under graduation level of the B. A. College of Agriculture, Anand Agricultural University, Anand (Gujarat). This college was selected purposively because of the following reasons:

1. This college is the oldest agricultural college of the state.
2. The students of this college have been getting computer and Internet facility since 1999.

3. All the students have been providing Internet facility free of cost at central library.
4. So far there is no any study has been carried out to develop scale to know computer anxiety and understand level of computer anxiety among under graduate students of this Agriculture College as well as to study the factors affecting to it.

10. SAMPLE SIZE

To employ developed scale, a random sample of total 100 undergraduate students studying the B.Sc. (Agri.) programme of B.A. College of Agriculture, Anand was selected to employ developed scale.

11. RESEARCH DESIGN

As the part of study was concerned to employ developed scale on sample students to understand its utility to measure attitude of the students towards computer anxiety/nervousness, the ex-post-facto research design was used. Kerlinger (1976) stated that ex-post-facto research design is worthy to apply when the independent variables have already acted up on.

12. DERIVATION OF HYPOTHESIS

On the basis of objective of the study, the following null hypothesis was formulated.

H_1- There will not be any relationship between various characteristics of the students and their degree of computer Anxiety/Nervousness.

13. METHOD AND DATA COLLECTION

The data were collected by personal interview from 100 randomly selected students. The respondents were contacted at their home, classes, college, hostel or library.

14. SELECTION OF THE VARIABLE

To employ developed scale, the variables were selected on the basis of extensive review of literature related to the subject and consultation with experts and finally the variables that were found to be most relevant to the present study are selected as under.

14.1 Independent Variables

Level of computer Anxiety/Nervousness.

14.2 Dependent Variables

1. Age
2. Academic performance
3. Father's education
4. Rural/Urban background (Native place)
5. Type of family
6. Wish to migrate to abroad
7. Wish to get higher academic degree
8. Possession of computer at home
9. Library exposure

14.3 Measurement of Variables

1. Age

Age of the respondents was measured in completion of year of the respondents at the time of interview. The respondents were classified in to two groups, viz., Up to 23 years and above 23 years.

2. Academic Performance

Academic performance of the respondents was measured in term of percentages or Over All Grade Points (OGPA) obtained by the students in their last semester or courses. To quantify this variable, the scores of 4, 3, 2 and 1 were assigned to the students with distinction, first class, second class and pass class, respectively.

3. Father's Education

Level of education of the father of respondents was measured in number of formal education in year completed by father of the respondents. For each completed formal year one score was assigned and for illiteracy '0' score was assigned. The respondents were classified in to five groups; Respondents with no any formal education of father, Respondents with education of father up to primary, Respondents with education of father up to S.S.C, H.S.C. and respondents with education of father upto college and above.

4. Native places of the Respondents

Looking to the native places of the respondents they were classified in to two groups and scoring was done as mentioned below:

1. Respondents with Rural background – 1 Score
2. Respondents with Urban background – 2 Scores

5. Type of the Family

Looking to the type of the family respondents were classified in to following groups and scoring was done as mentioned below:

1. Respondents with Joint family- 2 Scores
2. Respondents with Nuclear family – 1 Score

6. Wish to Migrate to Abroad

Looking to the wish to go abroad, the respondents were classified in to two following groups viz; Respondents with wish to go abroad–1 Score and respondents with no wish to go abroad – 0 Score.

7. Wish to get Higher Academic Degree

Looking to the wish to get higher or next academic degree, the respondents were classified in to two groups and scoring was done as respondents with a wish to have higher academic degree–1 Score and respondents without a wish to have higher academic degree – 0 Score.

8. Possession of Computer at Home

Looking to the possession of computer, the respondents were classified in to two groups and scoring was done as respondents with possession of computer at home – 1 score and respondents with no possession of computer at home – 0 Score.

9. Possession of E-mail ID

Looking to the possession of e-mail ID, the respondents were classified in to two groups and scoring was done as respondents with e-mail ID – 1 score and respondents with out with e-mail ID – 0 Score.

10. Library Exposure

Library exposure of the respondents and scoring procedure were done as per their frequency of the exposure viz. everyday exposure (4 score) , twice in a week exposure (3 score), once in fortnight exposure

(2 score), once in a month exposure (1 score) the never exposure (0 score).

14.4 Statistical Framework for Analysis of Data

The data were analyzed in order to make the findings meaningful for interpretation and drawing conclusions. For this, different statistical methods were used. Some of data were subjected to analyze in terms of percentage and frequency whenever necessary. At some other places, correlation of coefficient 'r' value was used to find out relationship between various characteristics of the students and their level of computer anxiety/ nervousness.

15. RESULT AND DISCUSSION

This part of the study deals with the procedure and methods adopted for the analysis and final selection of the items for the scale. First discussion has been given on final selection of the statement.

15.1 Final Selection of the Statement

The data from the 50 judges were arranged in the form as shown in following Tables 1 and 2. Table 1 shows the frequency distribution of judgments made by the judges for the statement No 5 on five categories.

Table 1: Frequency of the distribution of judgment made by judges on five categories

Sr. No.	*Distribution of the judgment for the statement No 5*	*Frequency*
1	IIII IIII IIII IIII IIII IIII IIII IIII IIIII	37
2	IIII IIII	08
3	I	01
4	IIII	04
5	Nil	00
Total		50

Table 2 : Summery of judgments made by judges on five categories for statement No. 5

Statement No. 5	*Sorting categories*					*Scale value*	*Q Value*
	1	*2*	*3*	*4*	*5*		
f	37	08	01	04	00		
p	0.74	0.16	0.02	0.08	00	1.18	0.73
cp	0.74	0.90	0.92	1.00	1.00		

As shown in Table 2, three rows are used for each statement. The first raw gives the frequency (f) with which the statement was placed in each of the five categories. The second row gives these frequencies as proportions (p). The proportions are obtained by dividing each frequency by N i.e. the total number of the judges (here it is 50). The third row gives the cumulative proportions (cp), that is the proportion of the judgments in a given categories plus the sum of all the proportions below the categories.

If the median of the distribution of the judgment for each statement is taken as the scale value of the statement, than the scale values can be found from the data arranged in the number of the Table-2 by means of the following formula.

$$S = L + \frac{0.50 - \sum Pb}{PW \times i}$$

Substituting in the above formula to find out the value for the statement number 5 in Table 2, we have....

$$S = 0.5 + \frac{0.50 - 0.00}{0.74} \times i$$

$$= 0.5 + 0.78$$

$$= 1.18$$

(The interval represented by the number assigned to the given category is assumed to range from 0.5 of a unit below to 0.50 of a unit above the assigned number. Thus lower limit of the interval represented by the category assigned the number 1 is 0.5 and the upper limit is 1.5)

The scale value can be found in the same manner for the other statements.

Thurstone and Chave (Edwards, 1957) used the inter-quartile range Q as a means of the variation of the distribution of the judgments for a given statement. To determine value of Q, we need to find two other point measures, the 75th centile and 25th centile. The 25th centile can be obtained by the following formula.

$$C_{25} = L + \frac{\sum Pb}{PW} \times i$$

Where,

C_{25} = The medium or scale value of the statement

L = The Lower limit of the interval in which the 25th centile falls

Pb = The sum of the proportion below the interval in which the 25th centile falls

Pw = The proportion within the interval in which the 25th centile falls

i = The width of the interval and is assumed to be equal to 1.0 (one).

For the statement number 5 in Table 2 we have .....

$$C_{25} = L + \frac{0.25 - Pb}{PW} \times i$$

$$C_{25} = 0.5 + \frac{0.25 - 0.00}{0.74} \times i$$

$$= 0.5 + 0.33$$

$$= 0.83$$

The 75th centile was obtained by the following formula.

$$C_{75} = L + 0.25 - \frac{\sum Pb}{PW} \times i$$

Where,

C_{75} = The medium or scale value of the statement

L = The Lower limit of the interval in which the 75th centile falls

Pb = The sum of the proportion below the interval in which the 75th centile falls

Pw = The proportion within the interval in which the 75th centile falls

i = The width of the interval and is assumed to be equal to 1.0 (one).

For the statement number 2 in table 2 we have

$$C_{75} = L + 0.75 - \frac{\Sigma Pb}{Pw} \times i$$

$$C_{75} = 1.5 + \frac{0.75 - 0.74}{0.16} \times i$$

$$= 1.5 + 0.06$$

$$= 1.56$$

Then the interquartile range would be given by taking the difference between C_{75} and C_{25}, thus,

$$Q = C_{75} - C_{25}$$

Substituting the values

$$Q = 1.56 - 0.83$$

$$= 0.73$$

In this manner the interquartile range (Q) for each statement was worked out for determinations of ambiguity involved in the statements. Only those statements were selected whose median values were greater than Q value. In case of this statement, value S = 1.18 and Q= 0.73. Hence the statement number five was selected.

Thurstone and Chave (Edwards, 1957) described another criteria in addition to Q as a basis for rejecting statement in scales constructed by the method of the equal appearing interval. Accordingly when a few items had the same scale values, the item having lowest Q Values were selected. To understand this point we can examine some of the following statements of the scale given in Appendix 2.

Table 3: Method of selecting the statements for the scale based on Scale value and Interquartile range

Statement Number	*Scale Value*	*Interquartile range Q*	*Selected/Not selected*
21	1.00	1.23	Not selected
4	1.00	0.95	Selected
19	1.13	0.63	Selected
5	1.18	0.73	Selected
20	1.21	0.34	Selected
17	1.30	1.12	Selected
7	1.30	1.13	Not selected
10	1.90	0.92	Not selected
9	1.90	0.69	Selected
6	1.97	1.27	Selected
15	2.20	2.05	Selected
8	2.58	1.33	Selected
14	2.90	1.04	Not selected
13	2.90	0.43	Selected
11	3.00	0.90	Selected
2	3.60	2.31	Not selected
22	3.60	1.78	Selected
18	3.71	1.06	Selected
23	3.80	0.76	Not selected
24	3.80	0.75	Selected
16	3.89	0.61	Selected
1	4.60	2.03	Not selected
12	4.60	1.34	Not selected
3	4.60	0.58	Selected

15.1.1 Final Statement for Attitude Scale

When there was a good agreement among the judges, in judging the degree of agreement or disagreement of a statement, Q was small compared to the value obtained, when there was relatively little agreement among the judges. Only those items were selected whose median (scale) values were greater than Q values. However, when a few items had the same scale values, items having lowest Q value were selected. Based on the median and Q values 16 statements numbering 3, 4, 5, 6, 8, 9, 11, 13, 15, 16, 17, 18, 19, 20, 22 and 24 of the original list were finally selected to constitute attitude scale. The scale values range from 1.00 to 4.70 with 0.5 class intervals.

15.1.2 Method of Scoring

The selected 16 statements for the final format of the scale were randomly arranged to avoid response biases, which might contribute to low reliability and detraction from validity of the scale. Out of the 16 selected statements, eight statements were the favourable indicators of the unfavourable computer anxiety and eight statements were the indicators of computer anxiety. Against these 16 statements, there were five columns representing five points continuum of agreement and disagreement to the statements as followed by Likert (1932) in his summated rating technique of attitude measurement. The five points on continuum were strongly agree, agree, undecided, disagree and strongly disagree with respective weights of 5, 4, 3, 2, and 1 for the favourable statements and with the respective weights of 1, 2, 3, 4 and 5 for the unfavourable statements. The weights of Likert's technique and the scale value of Thurstone's technique were combined in the form of a product and the total score for an individual was the sum of the product.

The lowest score would be 16 and highest score would be 80, the higher score indicates that person has less degree of computer anxiety/ nervousness and the lower score indicates that person has high degree of computer anxiety/ nervousness.

15.1.3 Reliability of the Scale

A scale is reliable when it will consistently produce the same results when applied to the same sample. In the present study due to limited time and resources available to the author, only split-half method of testing reliability was used.

15.1.4 Spilt-half Reliability

This test is a measure of internal consistency of a scale. One application of the scale is sufficient to secure the measure, since treats two halves of one scale as if they were two separate scales. The scale can split into two equal halves by odd-even method. The odd even method is favoured because:

1) It assured parallelism so far as difficulty was concerned, because items were arranged in order of increasingly difficult.
2) It assured that approximately the same amount of time was devoted each half.
3) It treated to keep testing conditions more nearly constant for the two halves, for it was highly unlikely that conditions either external or internal to him would fluctuate systematically with alternating items.
4) It was safer than comparing the first against the second half, since differential information fatigue or cumulative item effect might be lowered or revised than the true correlation (Koura and Singh, 1968).

The 16 statements were divided into two halves with 8 odd numbered in one-half and 8 even-numbered statements in the other. These were administered to 25 respondents. Each of the two sets of statements was treated as a separate scale and then these two sub-scales were correlated (table-3). The co-efficient of reliability was calculated by the Rulon's formula (Guilford, 1954), which came to 0.716.

Table 4 : Split-Half Reliability

S. No.	*Total Score of odd statements*	*Total Score of even statements*	*Differenced*	*d^2 $(x-y)^2$*	*Total Score of odd and even statements*	*t^2*
	(x)	(y)	(x-y)	$(x-y)^2$	t = [x+y]	
10	28	32	-4	16	60	3600
11	32	34	-2	04	66	4356
12	32	31	1	01	63	3969
13	30	34	-4	16	64	4096
14	30	31	-1	01	61	3721
15	33	33	0	00	66	4356
16	30	32	-2	04	62	3844
17	31	36	-5	25	67	4489
18	30	35	-5	25	65	4225
19	31	33	-2	04	64	4096
20	30	29	1	01	59	3481
21	28	30	-2	04	58	3364
22	29	32	-3	09	61	3721
23	30	32	-2	04	62	3844
24	32	34	-2	04	66	4356
25	30	31	-1	01	61	3721
			∑d	∑d2	∑t	∑t2
			-42	140	1570	98840

(A) $\Sigma d = -42$

$\Sigma d^2 = -1402$

$\Sigma t = 1570$

$t^2 = 98840$

(B) Reliability coefficients (calculated by Rulon's formula)

$$r_{tt} = 1 - \frac{\sigma^2 d}{\sigma^2}$$

r_{tt} = Coefficient of reliability

d = Differences between two score of respondents

$\sigma^2 d$ = Variances of those two difference $\sigma^2 d$

$\sigma^2 t$ = Variances of total score

n = Number of respondents

$$\sigma^2 d = \Sigma d^2 - \frac{\frac{(\Sigma d)^2}{n}}{n}$$

$$\sigma^2 t = \Sigma t^2 - \frac{\frac{(\Sigma t)^2}{n}}{n}$$

$$= 140 - \frac{\frac{(-42)^2}{25}}{25}$$

$$= 140 - \frac{70.56}{25}$$

$$= 2.7776$$

$$\sigma^2 t = 9.76$$

$$r_{tt} = 1 - \frac{2.7776}{9.76}$$

$$= 1 - 0.284$$

$$= 0.716$$

15.1.5 Content Validity of the Scale

The validity of the scale examined for content validity by determining how well content were selected by discussion with specialists, extension academicians, etc. thus, the present scale satisfied the content validity.

15.1.6 Final Format of the Scale

The total selected 16 statements, for the final format of the attitude scale were randomly arranged to avoid response biases, which might contribute to low reliability and detection from validity of the scale. Out of the 16 selected statements, eight statements were the indicators of the unfavourable computer anxiety and eight statements were the indicators of favourable computer anxiety. Against these 16 statements, there were five columns representing a five point continuum of agreement and disagreement to the statements as followed by Likert [1932] in his summated rating technique of attitude measurement. The five points on continuum were strongly agree, agree, undecided, disagree and strongly disagree. Final format has been presented in Table 5.

Table 5 : Final format of the scale with scoring

S. No.	*Statement*	*Strongly Agree*	*Agree*	*Un-decided*	*Dis-agree*	*Strongly Disagree*
1	I would like to use computers, If given the opportunity. (+)	5	4	3	2	1
2	I dislike working with computer machine that is smarter than me. (–)	1	2	3	4	5
3	The challenge of learning about computers is exciting. (+)	5	4	3	2	1
4	I have trouble in knowing the technical aspects of computers. (–)	1	2	3	4	5
5	I think I will not be able to learn a computer programming language. (+)	5	4	3	2	1
6	I hesitate to make use of computer for fear of making blunders that I cannot correct. (–)	1	2	3	4	5

contd...

7	I am confident that I can learn computer skills. (+)	5	4	3	2	1
8	I think that only masterminded person can make use of computers. (–)	1	2	3	4	5
9	Anyone can learn to use a computer if they are patient and motivated. (+)	5	4	3	2	1
10	I am worried that if I start using computer I will become dependent on it and be unable to do some work without it. (–)	1	2	3	4	5
11	I think that with time and practice I will be as comfortable working with computers as I am with papers or typewriter. (+)	5	4	3	2	1
12	I like to avoid computer use because it is unfamiliar and somewhat unapproachable to me. (–)	1	2	3	4	5
13	I feel that to get best result one should use computer in day-to-day life. (+)	5	4	3	2	1

contd...

14	I feel myself incompetent to work with computers. (–)	1	2	3	4	5
15	I feel computers are necessary tools in both educational and work settings. (+)	5	4	3	2	1
16	I feel tension to use computers (–)	1	2	3	4	5

15.2 Application of Computer Anxiety/Nervousness Scale on the Students

The knowledge of computer operating is an essential, especially for agricultural students. Information Technology has become an important part in teaching, research and education. In a golden era of information super highway the knowledge of computer is an essential for all. Without the knowledge of computer, the person will be designated as "outdated". The knowledge of computer and Agriculture Information technology will be proved as milestone for the future generation of human resource in Agriculture. Hence, to measure the computer anxiety/nervousness of agriculture student the developed scale was applied on the students of B.A. College of Agriculture, Anand Agricultural University Anand of Gujarat State. The data regarding the attitude of the students about computer anxiety/nervousness are presented in Table-6.

The result indicates that 42.00 per cent of the students had medium computer anxiety, followed by 39.00 per cent with highly and only 19.00 per cent with low computer anxiety.

Table 6: Distribution of the respondents according to their level of computer anxiety/nervousness

N = 100

S. No.	*Level of computer anxiety/ nervousness*	*Number*	*Per cent*
1	Low (Score above 57)	19	19.00
2	Medium (Score between 47 to 57)	42	42.00
3	Highly (Score below 47)	39	39.00
	Total	100	100.00

As it was mentioned in methodology that the persons having higher score will be considered with low level of anxiety/nervousness, at the same time person having low score will be considered with higher level of computer nervousness. As per these criteria, students were classified in to three groups as presented in table. The result says that majority (81.00 per cent) of the students studying in agriculture faculty had medium and high level of computer anxiety.

Thus, it can be concluded that there was medium to high level of computer nervousness in student of B.A. College of Agricultural. The probable reason for that might be the computer science is recently emerging science. It is affected by many characteristics of the students and their problem also. India is rural based economic country, in rural area the computer experience and exposure is poor as compared to urban areas. In addition to this students from rural area were not having own computer at home. Thus, due to lack of possession of computer by some of the urban and majority of the rural students at their own home, medium to high level of computer anxiety/nervousness was observed among majority of the students.

15.3 The Profile of the Students

The students having different level of computer anxiety/ nervousness, this might be due to difference in their personal, social, communicational, economical or psychological characteristics, which are related to the student's knowledge, adoption, exposure, experience and well acquainted with computers. Keeping in view the profile of the students have been studied and these phenomena, are discussed here.

15.3.1 Age

Table 7 : Distribution of the students according to their age

N = 100

S. No.	*Age group*	*Number*	*Per cent*
1	Up to 23 years	68	68.00
2	Above 23 years	32	32.00
Total		100	100.00

Age of the student is an important factor, which influences the computer anxiety/nervousness of the students. Looking to this age was studied and the data presented in Table 7. Table indicates that majority (68.00 per cent) of the students had up to 23 years of age and about 32.00 per cent of them had above 23 years of age. It can be concluded that majority of the respondent was from age group up to 23 years and only 32.00 per cent from age group above 23 years.

15.3.2 Academic Performance

Table 8: Distribution of the students according to their academic performance

N = 100

S. No.	*Academic performance*	*Number*	*Per cent*
1	Pass Class	20	20.00
2	Second Class	28	28.00
3	First Class	28	28.00
4	Distinction	24	24.00
Total		100	100.00

The academic performance shows the brilliancy of the students in their education. The computer anxiety/nervousness is also affected by student's brilliancy. The brilliant student may have different computer anxiety than average student. Data presented in Table-7 shows the respondent was belonged to all four classes. The students had equal proportion of second and first class (i.e. 28.00 per cent) followed by 24.00 per cent with distinction and only 20.00 per cent of them had pass class. It can be concluded from the results that more than half (56.00 per cent) of the students had first to second class level of academic performance.

15.3.3 Father's Education

Father's education is an importance for their children in proper guiding, motivating and directing for higher education. High level of education of the father helps students to create conductive situation that promotes him or her towards progress. Father with high level of education will have good understanding regarding various modern ways of advancement of his son or daughter. He motivates his children to make best use of all the means of development. Data presented in Table 9 indicates that half of father of the students had educated up to college level (51.00 per cent), Followed by secondary school 30.00 per cent), Higher Secondary school (16.00 per cent) and only 3.00 per cent of them were educated up to primary. It can be concluded that half of the father of the students had up to college level of education and no one was with illiterate father.

Table 9: Distribution of the students according to their Father's education

N = 100

S. No.	*Father's education*	*Number*	*Per cent*
1.	Illiterate	00	00.00
2.	Primary	03	03.00
3.	Secondary	30	30.00
4.	Higher Secondary	16	16.00
5.	College	51	51.00
Total		100	100.00

15.3.4 Native Place

Table 10: Distribution of the respondent according to their native place

N = 100

S. No.	*Native place*	*Number*	*Per cent*
1.	Rural	46	46.00
2.	Urban	54	54.00
Total		100	100.00

The data presented in Table 10 indicated the native place of the students. It is said that generally, urban youths and people utilize modern facilities more than rural people and youths. Any recent innovations are rapidly adopted in urban area than rural area due to so many reasons and factors. Here, the students with urban background were 54.00 per cent and with rural back ground were 46.00 per cent. Fragment of rural students in Agriculture College (46 per cent) is good signs but it needs to improve to make overall progress of rural and agricultural dominated country like India.

15.3.5 Type of Family

Table 11: Distribution of the students according to type of family

S. No.	*Type of family*	*Number*	*Per cent*
1	Joint	32	32.00
2	Nuclear	68	68.00
Total		100	100.00

The data presented in Table 11 indicate that majority (68.00 per cent) of the agricultural students were from nuclear family followed by (32.00 per cent) were from joint family. It means majority of the agricultural students had nuclear type of family.

12.3.6 Wish to Migrate Abroad

Table 12: Distribution of the students according to their wish to migrate abroad

N = 100

S. No.	*Wish to migrate abroad*	*Number*	*Per cent*
1.	Yes	77	77.00
2.	No	23	23.00
Total		100	100.00

At present the computer is becoming a part and parcel of the life. Computer has taken place in every walk of the life. Wish to migrate abroad is one of the important motivational forces which may affect intensity and degree of computer anxiety/nervousness. The data presented in table 12 indicate that more than three fourth of the students had wish to migrate abroad for their carrier advancement as well as for job purpose. Only 23.00 per cent of the students had no wish to go abroad. It means that the students had strong desire to go abroad for their bright future. The study shows that now Indian urban and rural youths have realized to settle in abroad. This might be due to less job opportunity in our country. To restrict our young generation to go abroad Government should create job opportunity to our agricultural students.

15.3.7 Wish to get Higher Degree

Table 13: Distribution of the students according to their wish to get higher degree

N = 100

S. No.	*Wish to get higher degree*	*Number*	*Per cent*
1	With wish	78	78.00
2	Without wish	22	22.00
Total		100	100.00

Desire of any individual motivates him or her to do something to active it. It is kind of motivational force, which influences an individual to involve himself leading towards success. Computer is the best mean to collect useful information regarding the various institutions and organizations offering various kinds of academic degrees and diplomas all over the world through internet facility. Thus in this study also wish of getting higher degree of students was analyzed and result is presented in Table 13.

The data presented in Table 13 shows that more than three forth (78.00 per cent) of the students had wish to get higher degree and only 22.00 per cent of the student had no wish to have next degree. It can be concluded from the study that major segments of the agriculture students had wish to have next higher degree.

15.3.8 Possession of Computer at Home

Possession of computer at home by agriculture students is a current need of the time. The students having own computer at home can get linked with world and able to get any information from the universe by means of various tools and techniques of information technology. Agri-informatics and bioinformatics are the emerging areas of the agriculture are playing an important role in agricultural research, teaching and education. Data presented in Table-14 shows that 60 per cent of the agricultural students did not possess computer at home, only 40 per cent of the students had their computer at home.

Table 14: Distribution of the students according to possession of computer at home

N = 100

S. No.	Possession of computer at home	Number	Per cent
1	Possession of computer	40	40.00
2	Not possessing the computer	60	60.00
Total		100	100.00

Data indicates that there is a need of technical assistance, motivation and encouragement to the students to have their own computer at home. This needs to be intensified by any way for overall development of the agricultural student. Availability of computer facility on community base at village level can be other alternative to provide computer to rural youths.

15.3.9 Possession of E-mail ID

Table 15: Distribution of the students according to their possession of E-mail ID

N = 100

S. No.	*Possession of E-mail ID*	*Number*	*Per cent*
1	Possession of E-mail ID	60	60.00
2	Not Possessing E-mail ID	40	40.00
Total		100	100.00

Possession of E-mail ID is good sign for overall development of the students. By having such facility, they can have a good touch with world. To get well acquainted with changing world through speedy communication, the e-mail ID is an essential asset. It was pleasant to note that majority (60.00 per cent) of the students of this study had their e-mail ID. Even though, still today 40.00 per cent of them did not have their e-mail ID. They need some motives for the same. It can be seen from the data that 60.00 per cent of the agriculture students of this study had their e-mail ID.

15.3.10 Library Exposure

Table 16: Distribution of the students according to their library exposure

N = 100

S. No.	*Library exposure*	*Number*	*Per cent*
1	Never	01	01.00
2	Once in a month	21	21.00
3	Once in a fortnight	49	49.00
4	Everyday	29	29.00
	Total	100	100.00

The result indicates that nearly half (49.00 per cent) of the students of agriculture faculty were utilizing library once in a fortnight, followed by 29.00 per cent utilizing everyday, 21.00 per cent utilizing once in a month and only 1.00 per cent of them did not utilize library at all, respectively.

The result shows that nearly half (49.00 per cent) of the students of agriculture faculty had library exposure once in a fortnight. Only 29.00 per cent of the students had everyday exposure of college library.

15.4 Relationship between Profile of the Students and their Computer Anxiety

Table 17: Relationship between attitude of the students towards computer anxiety/nervousness and their profiles

S. No.	*Name of variable*	*"r' value*	
1	Age	0.3931*	S
2	Academic performance	- 0.0966	NS
3	Father's education	- 0.0905	NS
4	Native place	- 0.1219	NS
5	Type of family	- 0.0358	NS
6	Wish to migrate aboard	- 0.0207	NS
7	Wish to get higher degree	- 0.1661	NS
8	Possession of computer at home	- 0.0967	NS
9	Possession of E-mail ID	0.1116	NS
10	Library exposure	0.1846	NS

* Significant at 5 % level of probability, S=Significant and NS = Non-significant

15.4.1 Age & Computer Anxiety/Nervousness

The data presented in Table 17 revealed that age was observed positively significant with computer anxiety. As explained in methodology adopted in scoring degree of computer anxiety/ nervousness, it was said that computer anxiety/nervousness will be higher when person gets low scores and same time anxiety/nervousness will be considered lower when person gets high scores. Thus it can be said from the results shown in table that students having more age had high score means low level of computer nervousness.

It is understood that students having more age will have more degree of exposure and experience of computer. Thus they will have less nervousness of computer. This might be the reason to have less computer nervousness among those students who had more age. Thus hall hypothesis (H_0) was rejected and above conclusion was drawn.

15.4.2 Academic Performance and Computer Anxiety

The data presented in Table 17 shows that negative correlationship between academic performance of the students and their level of computer nervousness. It means that person having low level of academic performance had high score of nervousness, which according to methodology of scoring indicates that he had low level of computer nervousness. Result says that academically less active students were more active in computer and had less computer nervousness. But this trend was not observed up to the level of significance. Null hypothesis in case of academic performances was rejected.

15.4.3 Father's Education & Computer Nervousness

The data presented in Table 17 revealed that there was negative co relationship between father's education and their computer nervousness. It means that the student having low level of father's education had high score of nervousness, which according to methodology of scoring indicating that he had low level of computer nervousness. Results say that students with lover level education of their father were found more active in computer and had less computer nervousness. But this trend was not up to the level of significance; hence the null hypothesis for this was rejected.

15.4.4 Native Place and Computer Nervousness

The relationship between native place of the students and computer nervousness was observed negatively non significant. It means that as compared to rural students the urban student had higher computer nervousness, but this difference was not up to the level of significance. According to method of scoring the person having rural back ground had more score means they had less computer nervousness and vice versa in case of urban background students. It is striking to note that the students having rural background where more active in computer exposure and they had low level of computer anxiety. This might be due to more hunger among rural students to know new system of computer technology. Thus the null hypothesis for native place was rejected and above conclusion was drawn.

15.4.5 Type of Family and Computer Nervousness

The result presented in Table 17 shows that type of family of the students was observed negatively non-significant with their computer nervousness. According to methodology adopted in scoring, it can be said that degree of computer nervousness of the students was lower among the student having nuclear family background as compared to the students having joint type of family background. In fact this tendency was not up to the level of significant. It was concluded that there was negatively non-significant relationship between type of family of agricultural students and their computer nervousness. Hence, null hypothesis was rejected and above conclusion was drawn.

15.4.6 Wish to Migrate abroad and Computer Anxiety/ Nervousness

The data presented in Table 17 show that relationship between wish to migrate abroad of students and computer nervousness was found negative. It means the student having desire to stay in India had low level of computer nervousness. Here it is fascinating to know that in spite of the wish to stay in India they took more interest in computer thus they had less computer nervousness. In other words students have realized that in Indian condition also they will be requiring knowledge of computer and affiliation with it to make progress. But again here also this trend was not that effective and significant, hence the null hypothesis was rejected and above conclusion was drawn.

15.4.7 Wish to get Higher Degree and Computer Anxiety/ Nervousness

The data presented in Table 17 show that relationship between wish to get higher degree of students and their computer anxiety was negative and non significant. It means that computer anxiety was observed low among those students who had no desire to get higher degree. Here also it was remarkable to note that students with no any expectations to get higher degree realized additional knowledge of computer as an additional qualification for their betterment. In fact this inclination was not up to the level of significance, thus null hypothesis in this case was rejected and concluded that relationship between wish to get higher degree of students and their computer anxiety was negative and non significant.

15.4.8 Possession of Computer at Home and Computer

Possession of computer at home by students will be an essential phenomenon for days to come. It is generally believed that the students having own computer at home will have more exposure of computer operating and less computer nervousness but result shown in this study shows that students without having computer at home had less computer nervousness. The result presented in Table 17 is supporting that there was high score means low computer anxiety among those students who had no any computer facility at their home. But as discussed earlier, this was not up to the level of importance, null hypothesis (H_0) for this case was also rejected and it can be concluded that the students having computer at home had low level of computer anxiety.

15.4.9 Possession of e-mail ID and Computer Nervousness/ Anxiety

The data presented in Table 17 reveal that the relationship between possession of e-mail ID and their computer nervousness was found positive non-significant. It means the student having possession of e-mail ID got high score which according to methodology was considered as persons with low level of computer nervousness. It may be due that the student having E-mail ID can run and easily operate the computer for receiving and sending the E-mail. Naturally, he had no fear of computer, which leads, towards low level of computer anxiety.

This observation was not up to the level of significance therefore, null hypothesis (H_0) was rejected and above conclusion was drawn.

15.4.10 Library Exposure and Computer Anxiety/Nervousness

Library exposure makes the students aware of changing scenario of the world through the use of available information technology and computer in library. Library is an important place to update the knowledge of the students. Generally it is seen that the students having high library exposure have good knowledge of all those components available in the library to collect information. Nowadays IT and computers have been made available as important parts of any modern library. Thus it can be said that students having high library exposure have good connections with computers that may lead them towards low level of computer anxiety/nervousness.

Data presented in Table 17 show that there was negatively non-significant relationship between library exposure and computer nervousness of the students. As per methodology of scoring of computer nervousness, person having high level of library exposure got high score means he had less computer anxiety/nervousness and person having less library exposure got less score means he had high computer anxiety/nervousness. Thus it can be said that there was low level of computer anxiety/nervousness among those students who had better exposure of library. In fact this trend was not up to the level of significance and hence. The null hypothesis incase of library exposure was rejected and above conclusion was drawn.

16. SUMMARY AND CONCLUSION

The study under report has great practical use for the extension agencies, scientists, teachers and research workers engaged in the field of agriculture extension.

The empirical findings of this study indicated to construct and standardize a scale to measure attitude of the students towards computer anxiety/nervousness. In developing the scale 24 statements about computer anxiety/nervousness were collected. The data from 50

judges were arranged, from which 16 statements were selected based on the median and Q values to constitute attitude scale. By split half reliability test internal consistency of a scale was measured which come to 0.716.

17. LEVEL OF COMPUTER ANXIETY

Majority (81.00 per cent) of the student had high to medium degree of favourable attitude towards computer anxiety/nervousness. Therefore extension agencies and teaching institutes should motivate the students and educate them for the importance and potentialities of computer in present world and its application in the field of agriculture and allied fields.

18. PROFILE OF THE STUDENTS

1. **Age:** Majority of the respondents (68.00 per cent) were from age group of up to 23 years followed by 32.00 per cent from them were from the age group of above 23 years.
2. **Academic performance:** Majority (52.00 per cent) of the students had first class or first class with distinction in their academic performance.
3. **Father's education :** More than half (57 per cent) of their father's education was up to higher secondary and college level.
4. **Native place :** More than half (54.00 per cent) of the students were found with urban background and 46.00 per cent of the students were found with rural background.
5. **Type of family :** Majority of the agriculture student (68.00 per cent) were from nuclear type of family system and only (32.00 per cent) of them were from joint type of family system.
6. **Wish to migrate aboard :** Aspiration to go aboard either for further study or for permanent setting was found in majority of the student (77.00 per cent).
7. **Wish to get higher degree :** Among the respondents a majority of the students (78.00 per cent) wish to get higher academic degree.
8. **Possession of computer at home :** Majority of the students (60.00 per cent) did not possess computer at home and only 40.00 per cent of the student had computer at home.

9. **Possession of e-mail ID :** Majority of the students (60.00 per cent) possessed e-mail ID and only 40.00 per cent of the student did not possess e-mail ID.
10. **Library exposure :** Only 29.00 per cent of the students were utilizing and visiting the library every day followed by 49.00 per cent once in fort night, 21.00 per cent once in a month and only 1.00 per cent had no library exposure at all.
11. **Relationship between profile of the students and their computer nervousness:** The computer nervousness was observed more among those students who had young age means studying in first or second years. The other variables such as academic performance, father's education, native place, type of family, wish to migrate abroad, wish to get higher degree, possession of computer at home, possession of E-mail/ID and library exposure did not have significant relationship with the computer anxiety/ nervousness of the students.

19. IMPLICATION

The study facilitated in constructing and standardizing a scale to measure attitude of the students towards computer anxiety/ nervousness, which will serve as a guidance for teachers, extension agencies, agriculture schools and colleges, scientists, researchers and policy makers to understand their affection with computers and improve their association with computers while planning and implementing programmes related to students' development activities. In addition to this, students' role in transferring technology, education and overall development can also be studied and improved with the help of this scale by knowing their existing involvement in Information Technology or computer application.

The study implies that there is a need to decrease students computer anxiety in order to achieve higher knowledge-based approach to computer education and overall agricultural information technology. Some important implications emerging from the findings of the study are below:

1. Special measures have to be taken to increase an interest of the students towards computer education, starting from primary education to university.

2. More Investment from government to hasten computer awareness as well as to decrease computer anxiety of the students are current needs of the time.
3. The students should be given training for experience, exposure and complete operation of computer programmes to break their computer nervousness.
4. The opportunities for productive employment should be enhanced by establishing agro-based industries, agri-informatic centers, agri-polyclinics, agri-business centers and community Internet centers.

20. SUGGESTIONS FOR FURTHER STUDY

1. Similar study may be conducted in other institutes, so that the results of the study can be strengthened.
2. Similar studies may be conducted periodically with large sample to determine the computer nervousness and to increase the awareness of computer applications in all fields.
3. Similar studies may be conducted on computer literate and illiterate students in agriculture, dairy and veterinary.

21. LITERATURE CITED

1. Adams, W. M. (1990). Green Development: Environment and Sustainability in the Third.
2. Andrew, Trotter (2002). Internet access has no impact on test score, study says, http://sc-net-terc.edu/do.efm/paper8509/show/use/set-tech.int.
3. Anonymous (2001). Approaches of choice - Integrated Pest Management (IPM) and Integrated Vector Management (IVM)http://www.who.int/docstore/water_sanitation_health/pesticides/ch5.htm.
4. Anonymous (2001). Instant messaging.
5. Anonymous (2002a). Use of information technology in agriculture http://www.india-infoline.com.bisc/itin.pdf.
6. Anonymous (2002b). How academic librarians can influence students' web-based information choices. http://www.5.ock.org.
7. Anonymous (2003a). Parents survey result. http://www.parent.umn. Edu / previousarchive.html.

8. Anonymous (2003b). Interactive education : Impact of the internet on learning and teaching http:// aftnn.org / blanket / site / http:// ubmail.edu / harsham / software /Index.htm.
9. Anonymous (2004). Computer anxiety and computer self-efficacy, http://www. educat.hu-berlin.de/mv/forschung/first.htm.
10. Anonymous (2004). E-commerce sector impact study http://Inter/ ee00464e.html.
11. Arneson, P. A. and J. E. Losey (1997). Integrated Pest Management – Lecture Notes.
12. Barbara, A. Walczak (1996). The quality and quantity of elementary student's use of the internet as a resource in science problem solving.
13. Beaumont, P. (1993). Pesticides, Policy and People. London: Pesticide Trust.
14. Brown, S.E. and Vician, C. (1997). "An Examination of the Relationship Between Computer Anxiety, Communication Apprehension and Student Experiences with Electronic Mail, http: // www. usm. maine. edu /~com/lindap~1.htm.
15. Burleigh, J. R. and Black, L. L. (2000). Supporting Farmers towards Safe Year-round Vegetables in Periurban Manila, http://www.ruaf. org/1-3/15-16.html.
16. Carson, R. (1962). Silent Spring. New York: Houghton Mifflin Company.
17. Catherine N. M. (2003). Challenges in Implementing Integrated Pest Management (IPM) Practices: Implications for Agricultural Extension, http://www. aged.tamu.edu /aiaee/2003/Munyua 469-479.pdf.
18. Chopra, K. (1993). Sustainability of agriculture. Indian Journal of Agricultural Economics, 48(3): 527-537 Cornell University.
19. Curtis, J. Bonk (2002). Online training in an online world.
20. David piercey (2000). Student's psychological well-being and the internet – A review and discussion of current research http:// dtp. epsh.net / projects / wwwise2.htm#6.
21. Deloughry, T.J. (1993). Researchers Say "Technophobia May Afflict Millions of Students", The Chronicle of Higher Education (28 April): A25-A26.

22. Drs. Douglas W. Green and Thomas O' Brien (2002). The internet's impact on teacher practice and classroom culture. http:// www. thejournal.com/ magazine / vault/a4081.cfm.
23. Edwards, C.A.; Thurston, H.D., & Janke, K. (1991). Integrated Pest Management for Sustainability in Developing Countries. In National Research Council. Toward Sustainability: A Plan for Collaborative Research in Agriculture and Natural Resource Management (pp.109-133). Washington D.C. USA: National Academy.
24. Eysenck, H. J. and Crown, S. (1949). An experimental study in opinion attitude methodology. International J. of Attitude Research. 3 : 47-86.
25. Funderburk, E. F. and Higley, L. G. (1994). Management of Arthropod Pests. In J. L. Hatfield, & D. L. Karlen (Eds.). Sustainable Agriculture Systems (pp.199-228). Boca Raton: Lewis Publishers.
26. George W. Norton (1994). Sustainability and participatory Integrated Pest Management, http : // www. ag. vt. Edu / ipmcrsp/ annrepts / annrep / PIPM.html.
27. Gressard, C.P. and Loyd, B.H. (1986). The Nature and Correlates of Computer Anxiety in College Students, Journal of Human Behaviour and Learning, 3:28-33.
28. Hakkinen, P. (1994). Changes in computer anxiety in a required computer course. Journal of Research on Computing in Education, 27 (2) : 141-153.
29. Hakkinen, P. (1994). Changes in computer anxiety in a required computer course. Journal of Research on Computing in Education, 27(2): 141-153.
30. Hamizatul, Maizan (1998). Kefahaman Mahasiswa tentang teknologi maklumat : satu kajian of universiti kebangahan Malasiya. {Undergrakeates' understanding of information technology: a study at university keb angsaan Malaysia] Unpub. referred from the article arfah yusof, http://www.interasia/org/malasiya/ hashim-yusof.html.
31. Herbert, D. A.; Sean, Malone and Randy, Shank (2002). Results of the Integrated Pest Management (IPM) Grower Survey and Demonstration Project to Determine IPM Adoption Strategies for Corn, Soybeans, and Small Grain in the Coastal Plain of Virginia, http://www.dcr.state. va.us / sw / docs / ipmresults.pdf.

32. Igbaria, M. and Chakrabarti, A. (1990). Computer Anxiety and Attitudes towards Microcomputer Use, Behaviour and Information Technology, 9 : 229-241.
33. Jessic Donn; Richard, S.; Christian E.; Egoc, K.; Zacharya, C.; Jaime, B. K.; and Thick, R. (2001). Attitudes and practices regarding relationship formation on the internet. http:// www. users. muohio. edu / sherma / apaol.htm.
34. Karkar, B. R. (1998). Impact of National Watershed Development Project for rainfed areas, on farmer's knowledge and adoption of rainfed agro technology. Ph.D. thesis (unpub.), Guj. Agril. Univ., Junagadh Campus.
35. Kerlinger, F. N. (1976). Foundation of behavioral research, New Delhi. Surjee Publication, pp. 198-204.
36. Kiss, A. and Meerman, F. (1991). Integrated Pest Management and African Agriculture. Washington D.C.: The World Bank.
37. Knausenberger, W.I.; Schaefers, G.A.; Cochrane, J. and Gebrekidan, B. (2001). Africa IPM.
38. Krishna, V. V.; Byju, N.G. and Tamizheniyan, S. Integrated pest management in Indian agriculture: a developing economy perspective (1996) http://ipmworld.umn.edu/chapters/Krishna.htm.
39. Krishnatray, Pradeep and Kulshrestha, Abhijita (2002). Internet use among, youth. http://www.mica-india.not/output/pk.pdf.
40. Lefter Daku; George W. Norton; Douglas G. Pfeiffer; Gregory C. Luther; Charles W. Pitts; Daniel B. Taylor; Josef Tedeschini and Rexhep Uka (2000). Farmers' knowledge and attitudes towards pesticide use and olive pest management practices in vlora, Albania: Abaseline survey, http://www.ag.vt.edu/ipmcrsp/papers/alb2000wp.pdf.
41. Leslie MacDonald (2003). Management Adoption in British Columbia (1998). Practices, A component of the State of Resources Survey http://www.agf.gov.bc.ca/cropprot/ipm_report.pdf.
42. Likert, R. A. (1932). A technical for measurement of attitude scale. Arch. Psycho. 140.
43. Manikrama, A (2002). Factors affecting the adoption of IPM by Farmers in the Dry Zone of Sri Lanka, http://www.siu.no /noradrap.nsf /0/db986ba 3af21b 633c1256c590032a305.

44. Marcoulides, G.A. (1988). The Relationship between Computer Anxiety & Computer Achievement, Journal of Educational Computing Research, 4 : 151-158.
45. Mathew, Joy and K. Nagi; Reddy (1989). A scale to measure the attitude of rural youth towards agriculture, Maha. J. of Ext. Edu., Vol. III, 203-206.
46. Morse, S.; McNamara, N.; Acholo, M.; Okwoli, B. (2000). Visions of Sustainability: Stakeholders, Change and Indicators. Aldershot, England: Ashgate Publishing Limited.
47. Necessary, J.R. and Parish, T.H. (1997). The Relationship between Computer Usage and Computer-Related Attitudes and Behaviours, Education, Spring 1996, v116 n3: 384-387, http : // www. usm. maine. Edu / ~com/lindap~1.htm.
48. Neil, Marriott, Neil, Selwyn and Pru, Marriott (1999). Accounting student use of information and communications technology in two UK universities.
49. Pam Dupin-Bryant (2002). Reducing Computer Anxiety in Adults Learning to Use Microcomputers, http://www.joe.org/joe/2002october/tt3.shtml.
50. Patel, M. C. (2004). Construction of scale to measure attitude of the farmers towards integrated pest management (IPM), an assignment submitted to YCMOU, Nashik.
51. Patel, N. R. (1991). Dynamics of adoption of new agricultural technology and consequences in watershed area of Banaskantha and Mahesana district of Gujarat State; Ph. D. Thesis (unpub.), Guj. Agril. Univ., Sardarkrushinagar.
52. Patel, N. R.; Pandya, D. N. and Patel, B. T. (1995). Techno-economic change among beneficiary farmers in watershed area, Maha J. Ext. Edn., 14 : 25-32.
53. Patel, R. C. (2000). A study of consequences of adoption of watershed management technology by the beneficiary farmers in watershed area of Kheda district of Gujarat state. Ph. D. Thesis (Unpub.) G.A.U., S.K.Nagar.
54. Patel, R.C. (2000). A study on the consequences of adoption of watershed management technology by the beneficiary farmers in watershed area of Kheda District of Gujarat State. Ph. D. Thesis (unpub.), Guj. Agril. Univ., Sardarkrushinagar.

55. Popat, M. N. (1990). The scale to measure the knowledge, attitude and skill of farmers about sprinkler irrigation system, Ph. D. Thesis. (Unpub.) G.A.U., S.K.Nagar.
56. Pretty, J. (2002). Agri-culture: Reconnecting People Land and Nature. London : Arthscan.
57. Pretty, J. N. (1998). Supportive Policies and practice for Scaling Up Sustainable Agriculture. In N. Röling & M.A.E. Wagemakers (Eds.) Facilitating Sustainable Agriculture (pp.23-45). Cambridge, U.K.: Cambridge University Press. Publications Ltd.
58. Rosen, L.D.; Sears, D.C. and Weil, M.M. (1987). Computer phobia, Behaviour Research Methods, Instruments and Computers, 19:167-179.
59. Suh Chong-Hyuk (1999). Integrated Pest Management (IPM) and Green Farming in the Asiatic Rice-Based small farm agricultural structure In-depth study, http://www.unescap.org/rural/doc/ipm2002/ch14.pdf.
60. Teresa, Y. N.; Jessame, F.; Michael, Romary, Simmona – Hoda and Katy Sullivan (2003). Literacy survey – 2003. Executive summary.
61. Thurnstone, L. L. and Chavel, E. G. (1928). The measurement of opinion J. of abnormal social psychology, 22 : 415-30.
62. Tseng, H.M.; Tiplady, B.; Macleod, H.; Wright, P. (1997). Computer Anxiety and the Assessment of Mood and Cognitive Function, http://www. york.ac.uk.
63. Uhm, K. (2002). Integrated Pest Management. Retrieved March 15, 2002, from World. London: Routledge.
64. Wang, K. A. (1932). Suggested criteria for writing attitude statements. J. of Social Psychology, 3 : 367-73.
65. Wlodkowski, R. J. (1993). Enhancing adult motivation to learn: A guide to improving instruction and increasing learner achievement. San Francisco: Jossey-Bass.
66. Yeap, S. L. (1998). Kegunaan internet di kalangan pelajar.http://www.interasia.org/malaysial/hashim.yusof.html.
67. Yee, M. C. (1998). Potensi urusniaga elekromik melaluli internet: satu kajian kes di universiti Malaysia sarwak [The potential of electronic commerce via internet a case study at university Malaysia, Sarawak] A Baccalaureate degree project puper. Department of communication, universiti kebangsaan Malaysia (Unpub.). Referred from the article on "Internet in Malaysia" by

Rahmah Hashim and Arfah yusof. http:// www. Interasia.org./ Malaysia/hashim-yusof-html.

68. Yeung, yau yuen; Chgng May Hung. May; So wing Mui Winnie and Tsang po Keug, Eric (1997). Using internet by education : Training student – teachers. http:// www. ied.edu.ht/has/webauth/ 4hkwsl.

Appendix – I

Please tick mark (√) your valuable response for each statement in appropriate column depending on relevancy to be considered it as one of the statements to measure "Computer Anxiety/nervousness of people".

S. No.	*Statement*	*Strongly Agree*	*Agree*	*Un-decided*	*Dis-agree*	*Strongly Disagree*
1	I am lack of confidence about my ability to interpret a computer.					
2	I am eager to use a computer in my job.					
3	I think I will not be able to learn a computer programming language.					
4	The challenge of learning about computers is exciting.					
5	I am confident that I can learn computer skills.					
6	Anyone can learn to use a computer if they are patient and motivated.					
7	Learning to operate computers is like learning any new skill.					
8	I am worried that if I start using computer I will become dependent on it and be unable to do some work without it.					

contd...

9	I am confident that with time and practice I will be as comfortable working with computers as I am with papers or typewriter.					
10	I think that I will be able to cope up with the advances happening in the field of computer.					
11	I dislike working with computer machine that is smarter than me.					
12	I feel hesitant about using computers.					
13	I have trouble in knowing the technical aspects of computers.					
14	It gives worry to me to think that I could cause big problem in the computer by hitting the wrong key.					
15	I hesitate to make use of computer for fear of making blunders that I cannot correct.					

contd...

16	I think that only masterminded person can make use of computers.					
17	I would like to use computers, If given the opportunity.					
18	I like to avoid computer use because it is unfamiliar and somewhat unapproachable to me.					
19	I feel computers are necessary tools in both educational and work settings.					
20	I feel that to get best result one should use computer in day-to-day life.					
21	I feel that use of computer makes person dull					
22	I feel myself incompetent to work with computers.					
23	Use of computer is not that much important because I feel that it is just a fashion.					
24	I feel tension to use computers					

(a) Selected Scale values and Q values

21	1.00	1.23
4	1.00	0.95
19	1.13	0.63
5	1.18	0.73
20	1.21	0.34
17	1.30	1.12
7	1.30	1.13
10	1.90	0.92
9	1.90	0.69
6	1.97	1.27
15	2.20	2.05
8	2.58	1.33
14	2.90	1.04
13	2.90	0.43
11	3.00	0.90
2	3.60	2.31
22	3.60	1.78
18	3.71	1.06
23	3.80	0.76
24	3.80	0.75
16	3.89	0.61
1	4.60	2.03
12	4.60	1.34
3	4.60	0.58

(b) Arranged values (selected statements)

3	4.60	0.58
4	1.00	0.95
5	1.18	0.73
6	1.97	1.27
8	2.58	1.33
9	1.90	0.69
11	3.00	0.90
13	2.90	0.43
15	2.20	2.05
16	3.89	0.61
17	1.30	1.12
18	3.71	1.06
19	1.13	0.63
20	1.21	0.34
22	3.60	1.78
24	3.80	0.75

(c) Selected statements (bold letters)

S. No.	*Statement*	*Strongly Agree*	*Agree*	*Un-decided*	*Dis-agree*	*Strongly Disagree*
1	I am lack of confidence about my ability to interpret a computer.					
2	I am eager to use a computer in my job.					
3	I think I will not be able to learn a computer programming language. (+)					

contd...

4	The challenge of learning about computers is exciting. (+)					
5	I am confident that I can learn computer skills. (+)					
6	Anyone can learn to use a computer if they are patient and motivated. (+)					
7	Learning to operate computers is like learning any new skill.					
8	I am worried that if I start using computer I will become dependent on it and be unable to do some work without it. (–)					
9	I think that with time and practice I will be as comfortable working with computers as I am with papers or typewriter. (+)					
10	I think that I will be able to cope up with the advances happening in the field of computer.					
11	I dislike working with computer machine that is smarter than me. (–)					

contd...

12	I feel hesitant about using computers.					
13	I have trouble in knowing the technical aspects of computers. (–)					
14	It gives worry to me to think that I could cause big problem in the computer by hitting the wrong key.					
15	I hesitate to make use of computer for fear of making blunders that I cannot correct. (–)					
16	I think that only masterminded person can make use of computers. (–)					
17	I would like to use computers, If given the opportunity. (+)					
18	I like to avoid computer use because it is unfamiliar and somewhat unapproachable to me. (–)					
19	I feel computers are necessary tools in both educational and work settings. (+)					
20	I feel that to get best result one should use computer in day-to-day life. (+)					

contd...

21	I feel that use of computer makes person dull					
22	I feel myself incompetent to work with computers. (–)					
23	Use of computer is not that much important because I feel that it is just a fashion.					
24	I feel tension to use computers (–)					

(d) Original statements

Sr. No.	*Statement*	*Strongly Agree*	*Agree*	*Un-decided*	*Dis-agree*	*Strongly Disagree*
1	I think I will not be able to learn a computer programming language. (+)					
2	The challenge of learning about computers is exciting. (+)					
3	I am confident that I can learn computer skills. (+)					
4	Anyone can learn to use a computer if they are patient and motivated.(+)					

contd...

5	I am worried that if I start using computer I will become dependent on it and be unable to do some work without it.(–)					
6	I think that with time and practice I will be as comfortable working with computers as I am with papers or typewriter. (+)					
7	I dislike working with computer machine that is smarter than me. (–)					
8	I have trouble in knowing the technical aspects of computers.(–)					
9	I hesitate to make use of computer for fear of making blunders that I cannot correct. (–)					
10	I think that only masterminded person can make use of computers.(–)					
11	I would like to use computers, If given the opportunity. (+)					

contd...

12	I like to avoid computer use because it is unfamiliar and somewhat unapproachable to me. (–)					
13	I feel computers are necessary tools in both educational and work settings. (+)					
14	I feel that to get best result one should use computer in day-to-day life. (+)					
15	I feel myself incompetent to work with computers. (–)					
16	I feel tension to use computers (–)					

(e) Final selected statements

Sr. No.	*Statement*	*Strongly Agree*	*Agree*	*Un-decided*	*Dis-agree*	*Strongly Disagree*
1	I would like to use computers, If given the opportunity.(+)					
2	I dislike working with computer machine that is smarter than me. (–)					
3	The challenge of learning about computers is exciting. (+)					

contd...

4	I have trouble in knowing the technical aspects of computers.(–)					
5	I think I will not be able to learn a computer programming language. (+)					
6	I hesitate to make use of computer for fear of making blunders that I cannot correct. (–)					
7	I am confident that I can learn computer skills. (+)					
8	I think that only masterminded person can make use of computers.(–)					
9	Anyone can learn to use a computer if they are patient and motivated.(+)					
10	I am worried that if I start using computer I will become dependent on it and be unable to do some work without it.(–)					
11	I think that with time and practice I will be as comfortable working with computers as I am with papers or typewriter. (+)					

contd...

12	I like to avoid computer use because it is unfamiliar and somewhat unapproachable to me. (–)					
13	I feel that to get best result one should use computer in day-to-day life. (+)					
14	I feel myself incompetent to work with computers. (–)					
15	I feel computers are necessary tools in both educational and work settings. (+)					
16	I feel tension to use computers (-)					

Appendix - II

Study on the computer Anxiety / nervousness of students

1. Name of the students : ______________________________
2. Age : __________ Year __________
3. Academic performance : ____________ OGPA / Percentage
4. Father's education : Illiterate/ Primary / Secondary / HSC / College Level
5. Native places of the respondents : Rural / Urban
6. Type of the family : Joint / Nuclear
7. Wish to migrate to abroad: Yes / No
8. Wish to get higher academic degree : Yes / No
9. Possession of computer at home : Yes / No
10. Possession of E-mail ID : Yes / No

11. Library exposure : everyday / twice in a week / once in fortnight / Once in a month / never exposure
12. Kindly give your responses for following statements :

Sr. No.	*Statement*	*Strongly Agree*	*Agree*	*Un-decided*	*Dis-agree*	*Strongly Disagree*
1	I would like to use computers, If given the opportunity. (+)					
2	I dislike working with computer machine that is smarter than me.(–)					
3	The challenge of learning about computers is exciting. (+)					
4	I have trouble in knowing the technical aspects of computers.(–)					
5	I think I will not be able to learn a computer programming language. (+)					
6	I hesitate to make use of computer for fear of making blunders that I cannot correct. (–)					
7	I am confident that I can learn computer skills. (+)					
8	I think that only masterminded person can make use of computers. (–)					

contd...

9	Anyone can learn to use a computer if they are patient and motivated. (+)					
10	I am worried that if I start using computer I will become dependent on it and be unable to do some work without it. (–)					
11	I think that with time and practice I will be as comfortable working with computers as I am with papers or typewriter. (+)					
12	I like to avoid computer use because it is unfamiliar and somewhat unapproachable to me. (–)					
13	I feel that to get best result one should use computer in day-to-day life. (+)					
14	I feel myself incompetent to work with computers. (–)					
15	I feel computers are necessary tools in both educational and work settings. (+)					
16	I feel tension to use computers (–)					

5 Role of Tribal Farmwomen in Home, Crop and Live Stock Management in Navsari District of Gujarat State

ABSTRACT

Farmwomen are the backbone of Indian agriculture. Growing food has been an interminable saga of her life. Like other rural women, tribal farmwomen also play an important role in agriculture. Farmwomen play vital role within home as housewives in managing the domestic affairs and they work as co-partners in the farming profession. No field operation is beyond the reach of women. They are at them best in sowing, transplanting, weeding, manuring, harvesting, winnowing, threshing, storing, marketing and rearing livestock etc. Besides this they are the manager to the household activities. They take important decisions in the home and outside the home. Scientific achievements and modernization are yet to make an impact on them. Keeping this fact in view the present investigation on role of tribal farmwomen in agriculture in Navsari district was undertaken with following objectives. (i) To study the socio-economic characteristics of the tribal farmwomen. (ii) To study the participation of the tribal farmwomen in agriculture, animal husbandry and household activities and the relationship between selected independent variables with crop and animal husbandry practices. (iii) To study the tribal farmwomen's involvement in decision making in farm management, animal husbandry and home management.

From this study it was observed that majority of the tribal farmwomen were observed that they had middle age group were primary educated had household, farming and animal husbandry as occupation, possessed small to medium size of herd, had marginal and small size of land holding, had 2 to 5 number of children, had all the three size of family, i.e. small, medium and large, lived in nuclear type of family, married at the age of 18 years and above.

Farmwomen participation in pre-sowing and sowing operations revealed that the highest respondents engaged with sowing followed by stubble collection, clode crushing, manuring and seedbed preparation. Incase of interculturing operations the participation of the farmwomen were observed the highest in weeding followed by gap filling, application of fertilizer, bird scaring, irrigation, bunding and hoeing with hand.

In harvesting and post harvesting operations, the highest participation was obtained in nipping / picking and threshing followed by harvesting, winnowing, storage, making threshing yard, bagging, packing and marketing of agriculture products. In animal husbandry practices the frequency of participation of farmwomen was seen the highest in cutting and bringing a fodder followed by compost making, watering, feeding, milking to animals, cleaning of cattleshed and so on.

Farmwomen took a self-decision for decoration of house (79.17 %) and selection and preparation of food (70.83 %) in case of home management. Farm management was dominated by husband decision and majority of the farm management decision was taken by their husbands, animal husbandry management was completely dominated by women's self decision.

The relationship between independent variables like age, education, herd size, land holding, family size and number of children of the respondents and their participation in crop husbandry was observed positively significant. Whereas the negative relationship was observed incase of occupation, type of family and age at marriage.

The relation between independent variables of the respondents and their participation in animal husbandry was found negative for all of the independent variables except type of family only.

Navsari district had recorded 34 percentage of agricultural labour of the total workers in the district. Normally agricultural labours constitute the majority of the working population in rural areas. Therefore the findings emerged out of the study of the selected talukas might well represent the role behaviour of rural women in farm management.

1. INTRODUCTION

Farmwomen are the backbone of Indian agriculture. Growing food has been an interminable saga of her life. Like other rural women tribal women also play an important role in agriculture. Even cultural arthropological literature suggests that agriculture is invention of women. Farming in India is mainly a family occupation. Most of the family members are acutely engaged in farming. At present when the farm technology is changing at faster speed a farmer has to adopt this in order to become a competitive and efficient farmer. The change in farming has increased manifold. The farming capabilities for taking timely and judicious decisions by the farm families have a direct bearing on the agricultural development in country. It is well known fact that decision is the heart of management. Much of success of farm families depends upon how well the family members develop skills in decision making.

In India as per 2001 census out of the total 496 million of rural female population 23.9 % are workers and about 87 % female workers are found in agriculture. There has been little realization about the contribution of women in the economic activities of a country. This was a position in most of counties till recently and India is no exception.

Farmwomen play vital role within home as housewives in managing the domestic affairs and they work as co-partners in the farming profession. No operation in field is beyond them. They are best in sowing, transplanting, weeding, manuring, harvesting, winnowing, threshing, storing, marketing and rearing livestock etc. Besides they are the manager to the household activities. They take important decision in home and outside the home. Scientific achievements and modernization are yet to make an impact on them.

It is to be noted that the perception of the place of women in rural economy of a country is largely conditioned by overall socio-economic and cultural situation prevailing there. It holds all the more true for a developing country like India.

From ancient times women have played a pivoted role in agriculture. In family the closest associate of a farmer is his wife. It is natural that farmers' wives can play a very effective role in improving the farming; they contribute one third labour force required for farming operations and allied enterprises. Women in villages spend for every minute of their working hours. Their hard lives start early in morning. They work in the house and then rushing off the work for 6-8 hours on the farm. In addition to participation in farm activities and the physical work, in joint families not only wives but also the mothers have a large part in decision-making. Women's contribution to the farm sector has largely been ignored and inadequately understood in our country.

It is said that if you educate a man, you educate an individual and if you educate a women, you educate a family. This old proverbs holds true today with their contribution and influence remaining unchallenged. Women in the present age are facing the most challenging situation of performing their roles in outside the home for social and economic development of the nation. Farm and home are inseparable in India.

Agriculture is a predominant sector, which provides employment in rural areas. The female population constitutes nearly half of the total population. It is a well recognised fact that more than 60 % of agricultural operations have been traditionally handled by women. In hilly areas where men migrate to agricultural operations including crop planning and marketing of produce. In other area men are reluctant to share control with women. The contribution of women to the farm sector has largely been ignored and inadequately understood in the Indian context.

Though women play different roles in their home activities as wives, as mothers and as homemakers, they also play a pivotal role in agriculture and livestock management. They still continued to share number of farm operations with men from early ages of invention of agriculture to the present day of modern agriculture.

Woman as leaders woman have been an integral part of social structure not only because of their importance in the perpetuation of human race but also by virtue of their significant contribution to socio-economic progress. The place of rural women in India in socio-economic and political spheres in more depressed than that of urban counter parts. Although the rural woman have been contributing significantly to the social and economic progress of the country. Their participation in development programmes is only as passive beneficiaries rather than active contributors.

Looking the significant role of tribal women in agriculture and allied activities, the study on rural woman's role in farm management was undertaken keeping in view of the increasing importance of involvement of rural women in agricultural production programmes. Studies in this field so far have exhibited a little concern on rural woman's role in Gujarat state. Hence this study was conducted with an objective of ascertaining the role expectations and role performance of rural women in farm management.

2. THE FEW WORDS FOR WOMEN

In ancient India, women have been as less than a divine status. Laxmi, Durga and Saraswati are the three great Goddesses of prosperity, power and wisdom, respectively. The ancient scriptures declare that god lives where women are worshipped. Surprisingly the Sanskrit word "ABALA" is routinely used with reference to woman as weaker woman. The term "ABALA" is made up to two parts A (without) and BALA (strength). Thus literally it means without strength. However, the term can notes weakness rather than absence of strength.

The terminology is close that Manu, who saw women as dependents whom, he advised to function under the guardianship of father before marriage of the husband after marriage and the sons in widowhood.

Gandhiji explained women as pillars of strength, pride and faith. Gandhiji had deep faith on the strength of women in national struggle for freedom as a powerful moral force in society and less concerned with improving their status. It should have formed the base of post independences planning for women, as partners got lost in constitutional designation of woman a weaker section. He once said "Women are the companion of man, gifted with equal mental capabilities. She has the right to participate in the minute details of the activities of man. She has the same right of freedom and liberty as man has. The woman is key co-partner in her husband's method of agriculture and animal husbandry for solving the economic and nutritional problems of India" (Devdas *et al.*, 1972).

3. EXPLANATION OF PROBLEM

The present investigation had been undertaken particularly on the role of tribal farmwomen in agriculture in Navsari district, such studies have not been conducted so far in Navsari district where the present study was conducted. Therefore, it was considered worthwhile that the study of the nature would be fruitful. After the primary survey, interview schedule was prepared and farmwomen were personally interviewed by the investigator.

Secondary data and other relevant information for the study was gathered from the following sources.

Reference books, reports, bulletins and periodicals on the subjects published by different authors, organizations, institutions and agencies.

Published papers, census data and other report from the government and other agencies.

Postgraduate thesis pertaining to similar or relevant studies conducted elsewhere.

In Navsari district among 5 tribal talukas, two talukas were selected randomly. Three villages from each taluka were selected. Total six villages were selected, from each village 20 tribal farmwomen were selected with the help of random sampling method. Thus the sample consisted of 120 tribal farmwomen. A schedule was prepared in light of objectives of the study and respondents were individually interviewed. The data were analysed in the light of objectives.

In this connection Jawaharlal Nehru once said: "In order to awaken the people. It is the women who have to be awakened. Once she is on move the family moves, the village moves the nation moves". (Hiranand and Kumar, 1980).

Therefore, this problem is planned to understand the role of tribal farmwomen in agriculture, home and livestock management in Navsari district of Gujarat state.

3.1 Scope of the Study

The present study will generate useful information on various aspects of involvement of tribal farmwomen in different farm and off farm activities and decision-making. This investigation is of great significance in creating data base fuller understanding of the existing situation and also for the course of action to be taken in future.

Providing an insight into a typical agrarian profile of tribal farmwomen, this study will suggest several implications to policy makers, planners, administrators, scientist and change agents for transforming an agricultural based rural society.

3.2 Limitations of the Study

Time, money and availability of skill, the main constraints. The present study has been restricted as under:

(i) The area of the study was limited to only two talukas of Navsari district of Gujarat state.
(ii) The study was confined only to one hundred and twenty tribal farmwomen of the selected area.
(iii) The study was based on individual's perception and expressed opinion.

However, proper care and considerate thought have been exercised in making the study as empirical and systematic as possible.

4. OBJECTIVES OF THE STUDY

The research study is designed with an objective to explore the multidimensional roles and participation of tribal farmwomen of the Navsari district of Gujarat state in general and with following objectives in particular.

(i) To study the Socio-economic characteristics of the tribal farmwomen.
(ii) To study the participation of the tribal farmwomen in agriculture, animal husbandry and household activities and the relationship between selected independent variables with crop and animal husbandry practices.
(iii) To study the tribal farmwomen's involvement in decision making for home management, farm management and animal husbandry practices.

5. REVIEW OF LITERATURE

Gasson (1980) reported that women have significant role in British agriculture by maintaining and reproducing the labour force, assisting in farm production, decision-making and business administration and helping to improve the quality of life in rural areas. Women have three ideal role types, distinguished by frequency of manual work, responsibility for farm enterprises, division of labour between husband and wife.

Rattin (1983) reported that women represent 46 % of the total agricultural population in France, 6 % of are farmers themselves, 46% are wives of farmers, 47 % are other family member and 1 % are analysed as are certain socio-demographic data the number of women running farms has gone up 1 % annually since 1975 and in 1981, 10 % of farms run by women.

Oxman (1983) aims to describe the characteristics of peasant women's organizations and degree of women's participation in them.

He described the role of women in the field of traditional agriculture, the emergence of agrarian organizations and intermittent participation of the peasant woman and her restricted political participation.

Saxena and Bhatnagar (1985) reported that in Rajasthan during peak season tribal women spent 14.66 hours on daily activities, out of which 8.24 hours were spent on agricultural work and 6.42 hours on home activities where as, non-tribal women spent 8.14 hours on farm activities and 7.61 hours on home activities totaling 15.75 hours per day.

Sisodiya (1985) indicated that job traditionally done by the farmwomen in the Chambal region in order of importance were mainly the harvesting, threshing, hoeing, weeding, winnowing and miscellaneous work which together accounted in 86.85 per cent and 79.08 per cent in 1982-83 and 1977-78, respectively.

Malkit and Sharma (1985) reported that dominance of female as the main caretaker of animals was prevalent in all regions but the maximum number was observed in backward region as compared to advance region.

Badiger and Rao (1985) found that the farmwomen participated independently more in home area, while their husband's independent participation was almost the same in both farm and home areas. Joint decisions were high in both areas, being somewhat greater in farm than in home areas.

Vagani (1987) revealed that during ordinary days, the farmwomen belonging to small, medium and big size land holding spent 2.00 hours, 3.00 hours and more than 3.00 hours respectively on their own farm.

Patel *et al.* (1987) reported that farmwomen participated in gap filling, harvesting, threshing, thinning and grain storage etc. They also reported that major activities performed by farmwomen in dairying were cutting and bringing green fodder, shed cleaning, bath to cattle, milking, watering and feeding to animal etc.

Lepcha (1987) reported that majority of the farmwomen made their own decision on storage of food grain management of their own subsidiary enterprises and to some extent on cropping pattern and marketing of farm produce, where as the decision on social participation and expenditure pattern were taken up jointly with their counter parts.

Singh (1989) studied the need to fully realize the enormous potentials of the tribal women by engaging her more actively in all spheres of agriculture. He stated that the role of tribal women in agricultural production and others in stressed. Benefits of developmental initiatives have always escaped women due to a variety of important sociological factors.

Vicente (1989) stated the role of women in agriculture is usually relegated to the status of farmer's wife or unskilled manual worker. Recent changes in agricultural sector leads to decrease the female population and increased interest in the role of women in agriculture. Women in most European countries plays equal role in agriculture and seen as her husband's helper.

Parekh (1990) reported that social participation had no relationship with the role performance of rural women in farm management. He also stated that family income had no relationship with role performance of rural women in farm management.

Kulkarni *et al.* (1990) stated that nearly 50.00 per cent of farmwomen had high knowledge regarding improved agricultural practices, whereas 42.00 and 8.00 per cent of the respondents had medium and low knowledge, respectively.

Kulkarni and Nandapurkar (1991) stated that very few rural women (15.00 per cent) were taking their own decision in respect of purchase of seeds, fertilizers, animal produce and selection of cropping pattern.

Uwakah *et al.* (1991) studied the role of women as farmers in Nigeria. The rural women were the more likely to be known as a full-time farmer. They reported that women have become increasingly involved in producing tree crops, food crops and livestock since the end of the Nigeria civil war in 1970.

Mistra (1991) studied the women of the nishing tribe of Arunachal Pradesh. The role of women in the nishing society and economy is interesting, both in its traditional aspect and in the changes that are taking place. Hunting and keeping watch over the semi-domesticated stock of mithusns were the main preoccupations of men, leaving almost the axis around which the nishing economic life revolved. The role of women was dominant in almost all the social, political, ritual and agriculture.

Bagwe (1991) reported that grassroots reality of village women's lives and seriously hamper efforts to develop supportive programmers and policy initiatives so as to effectively address critical issues related to women and poverty in Kokan, Bombay. Women try to support themselves and their households without much help from their men are also mentioned.

Nostrakis (1992) revealed that Sahelian women have remained largely ignored in development policy and project implementation. Research from the field of women in development is synthesised to detail the Socio-economic and context of the Sahelian women and to draw the strong link between her agricultural role and food product. Her conviction is subsequently used as a benchmark to measure the ways in which agricultural undermining her economic and perpetuating her subordinate decreased food production and increased dependency and environmental deterioration.

Salick (1992) reported the single women practice agriculture and opportunistically construct subsistence strategies differently than most tribal members. Crop density is high while diversity is lower because crops needing heavy labour are avoided.

Nostrakis (1992) reported the enormous roles of women to subsistence agricultural production in West Africa. The women were remained largely ignored in development policy and project implementation. The theoretical framework of dominant western models of agricultural development overlooks women and their well-being. He concluded that agricultural self-reliance should be prioritized in order to over come the current food crisis with active involvement of women.

Garcia (1992) studied the quantitative and qualitative contribution of women to family farms in Spain. A large scale survey carried out by the ministry of agriculture which involved 6203 women across 16 farming landscapes, it revealed information on activities which span the farming and domestic spheres and which tend to be left out of official statistics. Women's contribution is most acknowledged, women were working over 6 hours per day in farming, well above the national average. He noticed that the women's contribution is important in rural development.

Rossier (1992) reported that the average women spent 20 hours in a week on farm work, with the emphasis shifting from stable and field work towards bookkeeping and direct marketing of agricultural produce. The women on dairy farms and with fruit and vegetable crops spend more time on farm work, as old women on mountain farms. He also noticed that women were also involved so called male dominated tasks such as milking and tractor driving.

Chaudhary and Ganorkar (1993) indicated that majority of the farmwomen participated in stubble collection, cleaning of seeds, selection and treatment of seeds, sowing, weeding, drying, cleaning grains and storage of seeds. They further added that allied farm activities like feeding animals, collecting fodder for animals, cleaning animals and rearing calf are also carried out by the farmwomen.

Thakor and Patel (1993) observed that majority of the farmwomen needed training on methods of seed treatment, storage of farm produce, raising nursery, feeding animals, milking, preparation of milk products and artificial insemination.

Everaet (1994) reported that women have always occupied a central place in agriculture but this has sometimes led to their social and professional advancement being held back. Things are changing rapidly in Belgium, the number of women living off agriculture dropped by 28 % between 1980 and 1991. Women spend 81.7 % of their active time in farms. Mainly in the larger farms, around 31.5 % of the women are involved full time, others are part-time or having no involvement in the running of the farm.

Habib (1996) studied the role of women in agriculture in Pakistan and stated that the use of pesticides and the harmful impact of this use.

Rathore and Gaur (1996) studied in Rajasthan, the nature and extent of contribution of farmwomen in agriculture, comparing the time utilization pattern of males and females in farm operations in tribal and non-tribal areas. Rural women work 15-18 hours a day. Farmwomen work longer than men. Most of the farmwomen's activities were related to intercultural and post harvesting activities in both tribal and non-tribal villages. There was significant difference in home and farm activities between tribal and non-tribal women.

Nasreen *et al.* (1996) studied the role of farmwomen in decision making in various family affairs. Majority of the respondents felt that the husband should have the dominant role in important decisions in family matters. A very few respondents were not satisfied with their role as housewives. Independent variables such as age, education, caste, income and family type of the respondents had no significant effect on their decision making process.

Ogier (1998) reported that potential role of women in west Sumatra in a traditionally matrilinear society, belonging to an agro-forestry zone. The home gardens are traditionally prepared by the women. He reported that development of agro-forestry activities are traditionally female dominated in the world.

Sinn *et al.* (1999) reported the role of women in agricultural sector, especially as keepers of small livestock such as sheep and goats, greatly increases worldwide food security by improving the health and livelihood of individual families. They further stated that while women are the mainstay of small scale agriculture, farm labour force and day to day family subsistence. They have more difficulties than men in giving access to resources such as land and credit and other productivity enhancing inputs and services. The percentage of women who experience poverty is greater than men.

Diwan Yogita (2000) major roles performed by the tribal farmwomen in maize production activities were; ploughing through hoe, sowing of good quality of seeds, seed treatment thinning, carrying thinned plant at home, inter-culturing, weeding, preparation of FYM, uprooting diseased plants harvesting, binding of straw bundles, separation of grains threshing, winnowing, drying and storage of grains.

Antoniades and Papayiannis (2000) studied the role of women in the family farm of the mountain region of Cyprus and reveled that the women in the zone were aged, with 44% being over 63 years old. The vast majority (82.9%) were at the same time housewives and farmers and only 13% were fully or partly employed in off-farm jobs. Younger women had a higher incidence of off-farm jobs mainly due to higher level of education and improved opportunities to choose their own employment status.

Bisht *et al.* (2000) studied the pivotal role of women in the hills of Arah village in Uttar Pradesh, central Himalayas, India and reveled that the maximum energy (average of the four classes for the individual category of work) was expended by women on household activities (169 Mcal per person per year), followed by agricultural operations (148 Mcal per person per year) and animal care (112 Mcal per person per year). The energy spent on different functions was also calculated for each age class. Girls in the age range 5-15 and women of 56 years and above spent their maximum energy on household activities (227 Mcal per person per year and 258 Mcal per person per year, respectively), while women in the age ranges 16-35 and 36-55 expended the maximum energy on agricultural operations (313 Mcal and 246 Mcal per person per year respectively). Overall women in the age range 16 to 35 expended the most energy on average and those of 15 and under the least.

Geeta *et al.* (2000) have studied the role performance, knowledge and opinion level of panchayat women members and extracted that the majority of GP women members had medium role performance level (37%) followed by low (34%) and high (29%) performance levels, whereas the majority of TP women members had low (40%) role performance level and an equal percentage of medium and high performance level (30%). Approximately 67% of ZP women members had medium role performance level and 33% had high role performance level. A higher percentage of GP women members (41%) had low knowledge level while the majority of TP and ZP women members had high (40%) and medium (66.66%) knowledge levels, respectively. More favourable opinions towards their panchayat raj system were shown by 47, 30 and 33.33% of the GP, TP and ZP women members, respectively. The majority of the TP and ZP women members had the maximum role performance indices of 68.44 and 90.00 in the case of welfare programmes. GP members had maximum role performance index of 62.92 in the village improvement programmes, followed by welfare programmes (59.73), education (57.88) and agriculture and allied areas (55.58).

Fremont (2001) stated that women make up over one third of the agricultural workforce. The share of women remains stable in spite of the declining numbers in agriculture: 12 % of women who work do so full time 1 holding in 5 is managed by women.

Lohani and Khatrichhetri (2001) studied about the women in sustainable agriculture development and environment and revealed that the Women in the country have primary role in the farming system, providing 55-82 per cent labor required for overall agricultural production. Women are also the principal users of natural resources and they can play major roles in sustainable agricultural development and environment protection.

Qamar *et al.* (2002) studied the role of skilled and unskilled factory working women in the rural economy of Punjab and revealed that 29, 22% and 6% of the respondents were with middle, metric and intermediate level of education respectively whereas majority (34%) were educated up to primary level. Similarly more (54%) respondents were living in nuclear family. A vast majority had no decision making power while 80% of the respondents did not get any training before starting their job.

Firdous *et al.* (2002) studied the role and status of women in Baluch families and revealed that the majority of the respondents wanted to educate their children but were dissatisfied with family matters especially concerning marriage, education and share from their family's property. According to the majority of the respondents, women's rights in family life need protection.

Today women are approx. 50 % of the undergraduate population in animal science (AS) departments at the original land grant state universities, but racial minorities lag far behind, in part because the schools created under the 1890 legislation provided a diversion away from the state universities (Beck and Swanson, 2003).

It is suggested that the Samridhi Mahila processing co-operative society in Himachal Pradesh (SMCS) can empower the women and assist in economic development (Sharma *et al.*, 2003)

Mrinali *et al.* (2004) reported that among various demographic and economic factors, farm size, number of female household members, and women's wage rates were found to have played a positive role in determining the demand for total female labour. Level of education was found to have discouraged total female labour. On the other hand, level of education, farm size and family income were found to be positively related to the demand for hired female labour.

Khanduri *et al.* (2004) studied the role of females in household and recreational activities and the time they spent for these activities and reported that the women spent 16.08 hours/day for household and non-household activities, whereas men spent only 6.85 hours/day for these activities. Comparative data show that women devote 54% of their time each day for productive activities whereas men devote 65% of their time each day for non-productive activities.

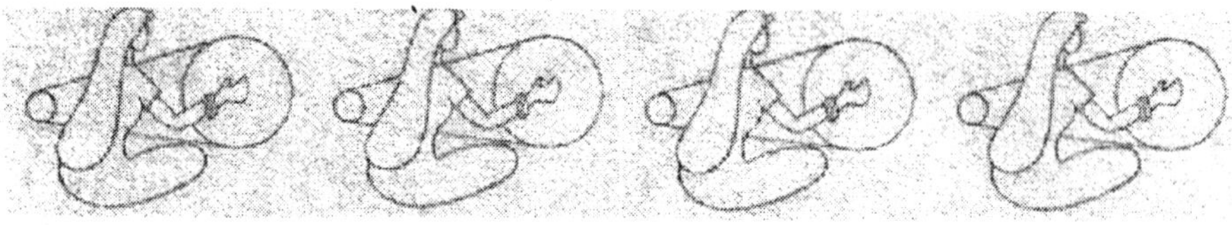

6. RESEARCH METHODOLOGY

The purpose of this section of the assignment is to describe the procedure followed for carrying out this study:

11.1 Plan of the study
11.2 Location of the study
11.3 Selection of the villages
11.4 Selection of the respondents
11.5 Construction of the interview schedule
11.6 Pre-testing of the schedule
11.7 Collection of the data
11.8 Research design
11.9 Derivation of hypothesis
11.10 Statistical tools and techniques used in analysis of data
11.11 Selection of the variables

6.1 Plan of the Study

The study was intended to call attention to the participation, awareness, training needs and constraints of the farmwomen in different faces of agriculture and home activities. Very few research studies have conducted on the contribution of women is in paucity.

Therefore, it was planned to investigate the woman's role and participation in agricultural activities.

6.2 Location of the Study

The presented investigation was carried out in the Navsari district of the Gujarat state, which is one of the tribal districts of the state. In selecting the district the main consideration was the agriculture as the main occupation of people living in such villages. Agriculture was the most important occupation of the people of the district and paddy was the main food crop produced. The other important crops grown are sugarcane, jowar, ragi, wheat, cotton, oilseed, pulses, vegetables and fruit crops viz., mango, banana and sapota.

6.3 Selection of the Villages

From the Navsari district, the six villages namely Adada, Aat, Chhapra, Hansapor, Matwad and Mogar were randomly selection for the study.

6.4 Selection of the Respondents

For the present study, from total selected 6 tribal dominated villages, 120 respondents were selected. To select respondents from the each village, help of officials and local people was taken. With the help of the list received from the officials a separate list of tribal farm families of each village was prepared. Further from the same list of tribal farm families, the tribal women who were decision makers and within the age group of 20-55 years were screened out. Finally by using random sampling technique, 20 respondents from each village were selected, thus a random sample of 120 respondents was selected for the study.

6.5 Construction of Interview Schedule

Due to high illiteracy among tribal women in the district, interview schedule was the only feasible method for gathering information from the respondents; firstly the structured interview schedule keeping in view the objectives of the study was prepared in English. The developed schedule was then translated in Gujarati for data collection from the respondents.

6.6 Pre-testing of Schedule

An interview schedule was pre-tested with ten tribal women of villages, which were not selected in the sample. Based on pre-testing, necessary modifications were made and the revised interview schedule was used for final data collection. An interview schedule was used for the final data collection. An interview schedule used for this study is appended.

6.7 Collection of the Data

Data were collected by arranging personal interview from the total selected 120 tribal farmwomen. The respondents were individually interviewed either at their homes or at their farms. At the time of interview, all possible efforts were made to develop rapport with them.

The aims and objectives of the study were explained to them with a view to facilitating them in giving correct response. The questions from the schedule were asked to them one by one and their responses were recorded in the schedule. The data were collected during the month of March 2005.

6.8 Research Design

As the study was concerned to find out the role performance of farmwomen in agriculture, ex-post-facto research design was used for this study. This design was used because the researcher did not have any control on the independent variables of the selected farmers.

Ex-post-facto research design is systematic experimental investigation in which control on independent variable is very weak and in this study almost no control was possible. Kerlinger (1976) stated that ex-post-facto research design is worthy to apply when the independent variables have already acted up on.

6.9 Derivation of Hypothesis

On the basis of objective of the study, the following null hypothesis was formulated.

H_0 – There will not be any relationship between selected characteristics of tribal farmwomen and their role in agriculture and animal husbandry.

6.10 Statistical Framework for the Analysis of Data

The following statistical tools were used for interpreting the data.

6.10.1 Number and Percentage

Simple comparisons were made on the basis of number and percentage.

6.10.2 Mean

The mean was obtained by the total number of score divided by total number of respondents.

$$\bar{X} = \frac{\Sigma xi}{n}$$

Where, $\bar{X}$ = Arithmetic mean of the sample

Σxi = The sum of all observation

n = Total number of observations in respective sample

6.10.3 Coefficient of Correlation

Coefficient of correlation was computed to find out the relationship between each of the independent variable and the dependent variable by employing following formula.

$$r = \frac{N\,\Sigma XY - \Sigma X\,\Sigma Y}{\sqrt{[\Sigma N \Sigma X^2 - (\Sigma X)^2]\,[\Sigma N(\Sigma y^2) - (\Sigma Y)^2]}}$$

Where, r = Correlation coefficient

X = Independent variable

Y = Dependent variable

N = Total number of respondents

6.11 Selection of the Variable

The variables understudy were selected on the basis of extensive review of literature related to the subject and consultation with experts and finally the variables that are found to be most relevant to the present study are selected as under.

6.11.1 Dependent Variable

Role of women in agriculture, animal husbandry and home management.

6.11.2 Independent Variable

Personal variable

1. Age
2. Education

Economical variable

1. Occupation
2. Herd size
3. Land holding

Social variable

1. Family size
2. Family type
3. Number of children
4. Age at marriage

6.11.3 Measurement of Variables

6.11.3.1 Measurement of Dependent Variables

6.11.3.1.1 Measurement of Role of Tribal Women in Different Activities

Role of tribal women in agriculture, animal husbandry and home management was measured with the help of their degree of involvement. It was measured in four point rating scale. They were asked to give their involvement in each activity in terms of "most frequent involvement", "frequently involvement", "rare involvement" and "no involvement" and scores of "4", "3", "2" and "1" were given respectively. To find out role in different activity, mean scores for each individual were worked out. To get total score of role of tribal women in agriculture and animal husbandry, scores gained by each individual for each sub items were added up.

6.11.3.1.2 Measurement of Decision making of Tribal Women in Different Activities

This was measured by collecting responses of the tribal farmwomen in four manners. They were asked to show their participation in decision making process related to various aspects of agriculture, animal husbandry and home management in terms of "Decision by herself", "Decision by her husband" "Joint decision by herself and any other member of her family including husband" "Decision not needed to take".

6.11.3.2 Measurement of Independent Variables

Age

Age in this study referred to the number of completed years of the respondents the respondents were grouped as under.

Young age	-	below 30 years
Middle age	-	between 31-50 years
Old age	-	above 51 years

Education

This was operationalised as whether the respondents can read and write. In this schedule information was asked about number of years of formal schooling completed.

Occupation

It was categorised as under .... Household + farming + agricultural labour household + farming + animal husbandry

Herd size

This was operationally defined as number of animals possessed by the respondent.

They were grouped as......

Not owning an animal

Owing up to 2 animals

Owing up to 3-5 animals

Owing above 5 animals

Number of children

The number of children of respondent were recorded and categorised as under.

No child

Up to 2 children

3 to 5 children

Above 5 children

Age at marriage

Respondents were asked for their age at the marriage and data were collected. It was categorised as under.

Age at the marriage up to 18 years

Age at the marriage above 18 years

Decision making

The respondents were asked about their participation in decision making related to farm management, home management and livestock management, decision making pattern were grouped into four groups.

Only self (women)

Only husband

Together with husband

Together with family members (joint decision)

Land holding

This refers to the number of hectares of land owned by respondent's husbands. In reality farmwomen were not the owner of land but the land possessed by their husband was taken into consideration. Based on land holding the respondents were categorised as......

Marginal	-	up to 1.0 hectare
Small	-	1.1 to 2.0 hectare
Medium	-	2.1 to 4.0 hectare
Large	-	above 4.0 hectare

Family size

Size of family was measured on the basis of total number of members in the family of respondents. On the basis of information collected into three groups.

Small family	(up to 5 members)
Medium family	(6 to 8 members)
Large family	(above 8 members)

Family type

The family types of respondents were grouped into two categories viz. nuclear type and Joint type.

7. RESULT AND DISCUSSION

This chapter highlights the findings of the investigation. The findings are furnished under the following sections.

12.1 Socio-economic characteristics of the tribal farmwomen

12.2 Participation of tribal farmwomen in home, crop and livestock management practices

12.3 Involvement of tribal farmwomen in decision-making process for farm management, home management and animal husbandry

12.4 The relationship between selected independent variables and crop husbandry

12.5 The relationship between selected independent variables and animal husbandry

7.1 Socio-economic Characteristics of the Tribal Farmwomen

This section deals with the distribution of the respondents according to their personal, socio-psychological and economic characteristics. The data collected, analysed and the results are presented in the following sections.

7.1.1 Age

Age of the women is an important factor, which influences the behaviour pattern. To study this factor the farmwomen were classified into three age groups, viz., young, middle and old age.

Table 1: Distribution of tribal farmwomen according to their age

N = 120

Sr. No.	*Age group*		*Number*	*Per cent*
1	Young age	(below 30 years)	52	43.33
2	Middle age	(30 to 50 years)	58	48.33
3	Old age	(above 50 years)	10	8.34
Total			120	100.00

The data presented in Table 1 indicate that nearly half (48.33 per cent) of the tribal farmwomen had middle age, followed by young age (43.33 per cent) and only 8.34 per cent with old age. It can be concluded that majority of the tribal farmwomen with young and middle age were engaged more in agricultural activities. The old age groups with poor health might have reduced their participation in agriculture production.

7.1.2 Education

It is evident from the data presented in Table 2 that major segment (50 per cent) of the tribal farmwomen were educated upto primary level, followed by 39.16 per cent with no any formal education, whereas only 7.50 per cent and 3.34 per cent of the tribal farm women were educated upto high school and higher secondary level and above higher secondary level, respectively.

Table 2: Distribution of tribal farmwomen according to their education

N = 120

Sr. No.	*Level of education*	*Number*	*Per cent*
1	Illiterate	47	39.16
2	Primary level	60	50.00
3	High school and higher secondary	09	7.50
4	Above higher secondary	04	3.34
Total		120	100.00

This might be due to their need to involve in economic activity from the childhood due to poor economic condition, less interest, understanding and realization regarding importance of education in the life. The tribal women are more orthodox and showing poor interest in taking formal education due to orthodox custom might be the another reason to have law level of education among them.

7.1.3 Occupation

The respondents were asked to furnish the information regarding their occupation and the data in this regard are presented in Table 3.

Data presented in Table 3 revealed that majority of the tribal farmwomen (89.17 per cent) had household + farming alongwith animal husbandry as their main occupation, followed by only 10.83 per cent with involvement in only agriculture and agriculture labour work as their source of livelihood. The results indicate that majority of respondents were found to be dependent on farming with animal husbandry.

Table 3: The tribal farmwomen according to their Occupation

N = 120

Sr. No.	*Occupation*	*Number*	*Per cent*
1	Household + farming + agriculture labour work	13	10.83
2	Household + farming + animal husbandry	107	89.17
Total		120	100.00

12.1.4 Herd Size

The respondents were asked to mention their herd size and the data available would be classified into four groups.

Table 4: Distribution of tribal farmwomen according to their herd size

N = 120

Sr. No.	*Hear size*	*Number*	*Per cent*
1	No animal	13	10.83
2	Small (upto 2 animals)	34	28.33
3	Medium (3 to 5 animals)	51	42.50
4	Large (above 5 animals)	22	18.34
Total		120	100.00

The data in Table 4 indicate that slightly more than two fifth of respondents (42.50 per cent) had medium herd size (3 to 5 animals) followed by 28.33 per cent with small and 18.34 per cent with large size of herd, respectively.

Thus, it can be said that majority of tribal farmwomen had 3 to 5 animals. Dairy is well developed and marketing is not a problem in this taluka. Thus, to earn additional income majority of the tribal farmwomen preferred to have animal husbandry with agriculture.

7.1.5 Land Holding

Data regarding distribution of farmwomen according to the size of land holding are presented in Table 5. Data indicate that majority (34.17 per cent) of the tribal farmwomen had marginal size of land holding, followed by 28.33 per cent with small, 22.50 per cent with medium and only 15.00 per cent had large or say above 4.00 ha. of land holding.

It can be concluded from above finding that majority of the respondents had small size of land holding.

Table 5: Classification of tribal farmwomen with land holding

N = 120

Sr. No.	*Size of land holding*	*Number*	*Per cent*
1	Marginal (up to 1 ha)	41	34.17
2	Small (1.1 to 2 ha)	34	28.33
3	Medium (2.1 to 4 ha)	27	22.50
4	Large (above 4 ha)	18	15.00
Total		120	100.00

7.1.6 Number of Children

Data collected in respect of number of children of the tribal farmwomen was presented in Table 6.

Table 6: The tribal farmwomen according to number of children

N = 120

Sr. No.	*No. of children*	*Number*	*Per cent*
1	No child	07	5.83
2	Up to 2 children	31	25.83
3	3 to 5 children	65	54.17
4	Above 5 children	17	14.17
Total		120	100.00

Data presented in Table 6 indicated that majority of the tribal farmwomen (54.17 per cent) had 3 to 5 number of children, followed 25.83 per cent with up to two, 14.17 per cent with above 5 and only 5.83 per cent of tribal farmwomen has no child. It can be concluded from this finding that majority of tribal farmwomen had 3 to 5 children. This might be due to lack of awareness and law level education.

7.1.7 Family Size

Table 7: Distribution of tribal farmwomen according to size of family

N = 120

Sr. No.	*Size of family*	*Number*	*Per cent*
1	Small (up to 5 members)	40	33.33
2	Medium (6 to 8 members)	63	52.50
3	Large (above 8 members)	17	14.17
Total		120	100.00

A close look to the table reveals that half of the tribal farmwomen (52.50 per cent) had medium size of family followed by 33.33 per cent with small size of family and only 14.17 per cent of them had large size of family. It can be concluded that majority of the farmwomen had medium and small size of family.

7.1.8 Family Type

There are two types of the family, nuclear and joint family. The data presented in Table 8 reveal that majority of the tribal farmwomen lived in nuclear type of family (63.33 per cent) and 36.67 per cent of tribal farmwomen lived in joint type of family. From this finding it can be concluded that majority of the tribal farmwomen lived in the nuclear type of family.

Table 8: Distribution of tribal farmwomen according to type of family

N = 120

S. No.	*Type of family*	*Number*	*Per cent*
1	Joint	44	36.67
2	Nuclear	76	63.33
	Total	120	100.00

7.1.9 Age at Marriage

Respondents were asked about their age of the marriage. Data regarding the age at the marriage are presented in the Table 9.

Table 9: The tribal farmwomen according to their age at marriage

N = 120

Sr. No.	*Age of marriage*	*Number*	*Per cent*
1	Up to 18 years	68	56.67
2	Above 18 years	52	43.33
	Total	120	100.00

It is evident from the table that more than half (56.67 per cent) of the tribal farmwomen were married at the age of 18 years, followed by 43.33 per cent get married after the age of 18 year.

7.2 Participation of Tribal Farmwomen in Home, Crop and Livestock Management Practices

7.2.1 Participation of Tribal Farmwomen in Agriculture

It is observed that when any person thinks and makes a picture of the farmers, he or she thinks always male as farmers but it is not true because many agricultural operations are performed more by the women. Keeping this in the mind, information about involvement of the tribal farmwomen in different operations was collected and results are presented in following tables.

7.2.1.1 Participation of Tribal Farmwomen in Pre-sowing and Sowing Operation

The perusal of the data in Table 10 revealed that the highest participation of tribal farmwomen was observed in sowing/transplanting followed by stubble collection, clode crushing, manuring and seedbed preparation / nursery.

Table 10: The role of tribal farmwomen in pre-sowing and sowing operation

N = 120

Sr. No.	*Pre-sowing and sowing operation*	*Mean value*	*Rank*
1	Sowing / transplanting	2.63	I
2	Stubble collection	2.51	II
3	Clode crushing	2.40	III
4	Manuring	2.02	IV
5	Seedbed preparation / nursery	1.32	V

7.2.2 Participation of the Tribal Farmwomen in Interculturing Operation

Frequency of participation by farmwomen in various interculturing operations was measured with the help of 4 point rating scale. Most frequently, frequently, least frequent and no participation with scale value of 4, 3, 2 and 1, respectively.

Table 11: Distribution of tribal farmwomen according to their participation in interculturing operation

N = 120

Sr. No.	*Interculturing operation*	*Mean value*	*Rank*
1	Weeding	3.16	I
2	Gap filling	2.22	II
3	Application of fertilizer	2.14	III
4	Bird scaring	1.69	IV
5	Irrigation	1.24	V
6	Bunding	1.09	VI
7	Hoeing with hand	1.07	VII

The data presented in Table 11 revealed that the highest participation of the tribal farmwomen was observed in weeding, followed by gap filling, application of fertilizer, bird scaring, irrigation, bunding and hoeing with hand, respectively.

7.2.3 Participation of Tribal Farmwomen in Harvesting and Post-harvesting Operation

Data collected from the respondents regarding frequency of participation in harvesting and post harvesting operations were given in Table 12.

Table 12: Participation of tribal farmwomen according to their participation in harvesting and post-harvesting operations

N = 120

S. No.	*Harvesting and post-harvesting operations*	*Mean value*	*Rank*
1	Nipping / picking and threshing	3.22	I
2	Harvesting	2.93	II
3	Winnowing	2.34	III
4	Storage	2.12	IV
5	Making threshing yard	1.93	V
6	Bagging / packing	1.54	VI
7	Marketing of agricultural produce	1.29	VII

Perusal of the data presented in Table 12 indicated that the highest participation of the tribal farmwomen was observed in nipping / picking and threshing followed by harvesting, winnowing, storage, making threshing yard, bagging / packing and marketing of agricultural produce, respectively. The roles of tribal farmwomen in all the operations were observed up to the importance level.

7.2.4 Participation of the Tribal Farmwomen in Animal Husbandry Practices

Animal husbandry is one of the important occupations preferred to have along with agriculture by majority of the rural people. It is said that as compared to male members of farmers family, women play significant role in performing most of the roles of animal keeping. To understand real picture about tribal family, data were collected and results are presented in Table 13 and they are explained in following paragraphs.

The highest mean score of participation of tribal farmwomen was obtained incase of cutting and bringing a fodder. The next animal husbandry related operation performed by tribal farmwomen was compost making followed by watering and feeding to animals, milking, cleaning of cattleshed, selling of milk and its product, bathing of animals, preparation of milk products, grazing of animals, taking animal for bull and veterinary services. It can be concluded that the role of tribal farmwomen in most of the animal husbandry related operations were seen important. The possible reason for higher participation of women in animal husbandry would be that the most of the tribal farmwomen were possessing more than 2 animals. Agriculture and animal husbandry are the two sides of the same coin. The economy of the farming communities is basal on agriculture as well as on animal husbandry.

Table 13: Participation of the tribal farmwomen according to their participation in animal husbandry practices

N = 120

Sr. No.	*Animal husbandry operations*	*Mean value*	*Rank*
1	Cutting and bringing a fodder	3.62	I
2	Compost making	3.61	II
3	Watering and feeding to animals	3.40	III
4	Milking	3.22	IV
5	Clearing cattleshed	3.01	V
6	Selling of milk and its products	2.23	VI
7	Bathing animals	2.04	VII
8	Preparation of milk products	1.64	VIII
9	Grazing	1.63	IX
10	Taking animal for bull service	1.57	X
11	Taking animal for veterinary service	1.48	XI

Thus, to increase family income tribal farmwomen might have taken keen interest in most of the animal husbandry related operations. Other reason to have higher participation of tribal farmwomen in most of the animal keeping activities might be that animal husbandry related work has been giving predominantly to the female members of the family in rural areas by the male members from beginning.

7.3 Involvement of the Tribal Farmwomen in the Process of Decision Making of Farm Management, Mome Management and Animal Husbandry

In order to plan and carryout farm and home programmes successfully, there is need to know about decision-making process whether taken at individual level or jointly. Women as wives and mothers have a considerable part in taking decisions regarding farm and home. Decision-making is important because much of the success of farming depends upon how well the family makes decision. In joint families not only the wives but also the mothers have a large part in decision making. To determine the decision making pattern, the decision makers were categorised into four groups "Decision by herself", "Decision by her husband" "Joint decision by herself and any other member of her family including husband" "Decision not needed to take'.

7.3.1 Involvement of Tribal Farmwomen in the Process of Decision making about Home Management

The data collected from the tribal farmwomen regarding their involvement in the process of decision making about home management are presented in Table 14.

Data indicates that the self-decision by the tribal farmwomen was dominated in case of decoration of house (79.17 per cent) and selection and preparation of food (70.83 per cent). The husband dominated decision making was observed in case of house repair (79.17 per cent) followed by borrowing money for home management (77.50 per cent) repayment of loan (66.67 per cent) manner of sawing (62.50 per cent), Children's education (58.33 per cent), children's occupation (45.83 per cent) and construction of new house (25.00 per cent), respectively.

Joint decision was dominated in case of children's marriage (81.67 per cent) followed by selling and purchasing of ornaments (75.83 per cent), construction of new house (75.83 per cent) manner of saving (35.83 per cent), respectively.

It can be concluded that the tribal farmwomen were independently taking a decision for decoration of house and selection and preparation of food, while husband dominated decisions were matters of house repairing, borrowing money for house management, repayment of loan, manner of saving and purchase of household articles. Joint decision made by tribal farmwomen and any members of their family were children's marriage, selling and purchasing of ornaments, constructionof new house and manner of saving.

Table 14: Tribal farmwomen involvement in the process of decision making for home management

N = 120

Home management	*Only self*		*Husband*		*Alongwith family (Joint decision)*		*Not related*	
	Number	*Per cent*	*Number*	*Per cent*	*Number*	*Per cent*	*Number*	*Per cent*
Construction new house	00	00	30	25.00	85	70.83	05	4.17
Decoration of house	95	79.17	05	4.17	20	16.67	00	--
Children's education	02	1.67	70	58.33	40	33.33	08	6.67
Children's occupation	02	1.67	55	45.83	48	40.00	15	12.50
Children's marriage	03	2.50	10	8.33	98	81.67	09	7.50
House repair	05	4.17	95	79.17	20	16.67	--	--
Purchasing of household articles	08	6.67	62	51.67	50	41.67	--	--
Selling and purchasing of ornaments	06	5.00	23	19.17	91	75.83	--	--
Selection and preparation of food	85	70.83	05	4.17	30	25.00	--	--
Borrowing money for home management	07	5.83	93	77.50	20	16.67	--	--
Repayment of loan	04	3.33	80	66.67	12	10.00	24	20.00
Manner of saving	02	1.67	75	62.50	43	35.83	--	--

7.3.2 Tribal Farmwomen's Involvement in the Process of Decision making about the Farm Management

Data collected from the respondents regarding their involvement in the process of decision-making about farm management are given in Table 15.

Table 15:Tribal farmwomen involvement in the process of decision making for farm management

N = 120

Farm management	*Only self*		*Husband*		*Alongwith family (Joint decision)*		*Not related*	
	Number	*Per cent*	*Number*	*Per cent*	*Number*	*Per cent*	*Number*	*Per cent*
Deciding area to be sown under different crops	05	4.17	75	62.50	40	33.33	--	
Introduction of a new crop variety	04	3.33	99	82.50	17	14.17	--	
Selection of seed	03	2.50	78	65.00	39	32.50	--	
When to irrigate fields	01	0.83	105	87.50	06	5.00	08	6.67
Quantity and type of fertilizers used on the farm	--	--	102	85.00	12	10.00	06	5.00
Using plant protection measures	01	0.83	92	76.67	05	4.17	22	18.33
Hiring farm labours	--		25	20.83	80	66.67	15	12.50
Buying farm machinery / equipment	02	1.67	96	80.00	22	18.33	--	--
Installing oil engine and electric motor	--	--	80	66.67	40	33.33	--	--
Buying and selling of land	--	--	34	28.33	86	71.67	--	--
Borrowing money for farm operations	--	--	84	70.00	36	30.00	--	--
Selling of surplus farm produce	03	2.50	47	39.17	70	58.33	--	--

It is evident from the data that in majority of the decisions regarding farm management were husband dominated such as when to irrigate the fields (87.50 per cent). Quantity and type of fertilizers to be used in the farm (85.10 per cent), introduction of new crop variety (82.50 per cent), buying farm machinery / equipment (80.00 per cent), using plant protection measures (76.67 per cent), borrowing money for farm operation (70.00 per cent), installing oil engine, electric motor and pumps (66.67 per cent), selection of seed (65.00 per cent) deciding area to be sown under each crop (62.50 per cent) and sown on, respectively. The joint decision made by tribal farmwomen and other members of family were hiring farm laborers (66.67 per cent), buying and selling of land (71.67 per cent) and selling of surplus farm produce (58.33 per cent), respectively.

It can be concluded that the most of the farm decisions were made by husband of tribal farmwomen followed few decisions made jointly by the tribal farmwomen after discussion with any of their family members. It can be further concluded that tribal farmwomen had a recessive role in decision-making process regarding farm management.

7.3.3 *Farmwomen's Involvement in the Process of Decision making for Animal Husbandry*

Data collected from the respondents regarding their involvement in decision making for animal husbandry are presented in Table 16.

The results indicate that the selection of fodder and feed was dominantly decided by farmwomen (81.67 per cent) followed by sale of milk and its product (52.50 per cent). The decisions regarding sale and purchase of animal (65.00 per cent), selection of animal breed (60.00 per cent) and keeping size of herd (40.00 per cent) were taken jointly by family members.

Perusal of the data presented in table 15 concluded that the decision regarding animal husbandry was dominated by tribal farmwomen themselves as well as by joint decision with family members. The husband remained recessive in decision making about animal husbandry. Majority of the farmwomen were taking self-decision regarding fodder and marketing of milk and milk products. The joint decisions taken were selection of animal breed, sale and purchasing of animal as well keeping size of herd, some of the decisions were not needed to take by them because some of them had no animal or some of them were not selling milk and its products.

Table 16:Tribal farmwomen involvement in the process of decision making for animal husbandry

N = 120

Animal husbandry	*Only self*		*Husband*		*Alongwith family (Joint decision)*		*Not related*	
	Number	*Per cent*	*Number*	*Per cent*	*Number*	*Per cent*	*Number*	*Per cent*
Selection of animal breed	07	5.83	29	24.17	72	60.00	12	10.00
Selection of fodder and feed	98	81.67	04	3.33	06	5.00	12	10.00
Sale and purchase of animals	13	10.83	15	12.50	78	65.00	14	11.67
Sale and milk and its products	63	52.50	08	6.67	18	15.00	31	25.83
Keeping size of herd	42	35.00	15	12.50	48	40.00	15	12.50

7.4 The Relationship between Selected Independent Variables and Crop and Animal Husbandry

Table 17: Relationship between independent variable of the respondents and their participation in crop husbandry

N = 120

Sr. No.	Independent variables	Correlation coefficient ('r' value)
I.	**Personal variables**	
1	Age	0.8430 **
2	Education	0.7409 **
II.	**Economical variables**	
3	Occupation	- 0.5553 *
4	Herd size	0.7854 **
5	Land holding	0.7774 **
III.	**Social variables**	
6	Family size	0.8074 **
7	Type of family	- 0.6284 **
8	Number of children	0.4918 *
9	Age of marriage	- 0.8709 *

* Significant at 5 %, ** highly at significant at 1 %.

7.4.1 Age and Participation

Data presented in Table 17 indicated that age of the respondent was observed positively significant with their participation in crop husbandry. It means old aged tribal farmwomen were more actively participated in agricultural operations as compared to young aged farmwomen. The trend of relationship says that old aged farmwomen had higher participation in crop husbandry, means farmwomen's participation in agriculture was increasing with increasing their age. Thus the null hypothesis (H_0) in case of age was rejected.

7.4.2 Education and Participation in Crop Husbandry

Level of education of the tribal farmwomen was positively highly significant with their participation in crop husbandry. High level of education might have given opportunity to farmwomen to understand the value of agriculture in a household economy. The education makes them awared with agro-based economy of the rural area. These might be the reason to have positive relationship between education of farmwomen and their participation in crop husbandry. Hence the null hypothesis (H_0) in relation to education was rejected and it can be said that there was significant relationship between the education of the farmwomen and their participation in crop husbandry.

7.4.3 Occupation and Participation in Crop Husbandry

The result shown in Table 17 reflects that there was negative significant relationship between occupation of tribal farmwomen and their participation in crop husbandry. It indicated that the role of tribal farmwomen in a crop husbandry was observed more among those tribal farmwomen who were involving in less number of occupations or only in agriculture. It is natural that person who is dependent only in one occupation will always try to involve more in the same occupation to get maximum advantage through it. This might be the reason to have better participation in crop husbandry among those farmwomen who were dependent only on agriculture. Thus, the null hypothesis (H_0) in case of occupation was accepted.

7.4.4 Herd Size and Participation in Crop Husbandry

The results revealed that the size of herd owned by tribal farmwomen was observed positively significant with their participation in crop husbandry. It means that the higher participation in crop husbandry was observed among those tribal farmwomen who had bigger herd size. It is likely that tribal farmwomen who possess more number of milch animals will required more fodder; they need to do more work in the field. Same situation was also observed here and it was seen that to accomplish fodder requirement, more participation in crop husbandry was observed among those tribal farmwomen who had big size of herd. Hence this relationship was observed significantly positive. Here, the null hypothesis (H_0) was rejected in this case too.

7.4.5 Land Holding and Participation in Crop Husbandry

The result shown in Table 17 reflects that there was positive significant relationship between the size of land holding of tribal farmwomen and their participation in crop husbandry. It means that more participation in crop husbandry was observed among such tribal farmwomen who had bigger size of land holding. It is natural that when tribal farmwomen possess more land holding, their participation will be more in crop husbandry to achieve better yield, higher profit and higher standard of living. Thus, tribal farmwomen with big size of land holding were more active in participating agricultural operations. It means that tribal farmwomen with small size of land holding might have preferred to have animal husbandry as additional source of income, in the situation they might preferred more in animals husbandry related activities to earn more income from the available small size of land holding. Hence, the null hypothesis (H_0) in relation to land holding was rejected.

7.4.6 Family Size and Participation in Crop Husbandry

Here, also positive significant relationship was observed between family size and the participation of tribal farmwomen in crop husbandry.

It is apparent that with increase in family size, the availability of tribal farmwomen for crop husbandry was observed more. In this study it was seen that majority of the tribal farmwomen had either 3 or more than 3 children. In case of the more children, there are chances for women member to engage other girl children in household activities and can participate herself more in agriculture to get higher production to fulfill the requirement of the family by means of agriculture as a source of livelihood. Hence, the null hypothesis (Ho) in this case was rejected.

7.4.7 Type of Family and Participation in Crop Husbandry

Data presented in Table 17 show that there was negative significant relationship between type of family of the tribal farmwomen and their participation in crop husbandry activities. It means participation in crop husbandry was seen higher in case of tribal farmwomen with nuclear type of family and vice versa for joint type of family. Hence, the null hypothesis (Ho) in case of type of family was rejected and above conclusion was drawn.

7.4.8 Number of Children and their Participation in Crop Husbandry

Data presented in Table 17 revealed that number of children and their participation in crop husbandry was seen positive significant. It means the tribal farmwomen who had more number of children, had more participation in crop husbandry. The tribal farmwomen having more number of children had more time for field operations as the household responsibilities can be handed over to the children. Another probable reason might be that with increasing the number of children the responsibility of tribal farmwomen increases and they need to earn more by sound crop husbandry with higher production. Thus tribal farmwomen with more number of children had higher participation in crop husbandry as compared to tribal farmwomen with less number of children. Here, also the null hypothesis (H_0) was rejected.

7.4.9 Age at Marriage and their Participation in Crop Husbandry

The result shown in Table 17 reflects that there was negative significant relationship between age of tribal farmwomen at marriage and their participation in crop husbandry. It means tribal farmwomen married at young age had more participation in crop husbandry as compared to tribal farmwomen married at old age. Hence, the null hypothesis (H_0) in this case is rejected.

12.5 Relationship between Independent Variables of the Respondents and their Participation in Animal Husbandry

Table 18: Relationship between independent variable of the respondents and their participation in animal husbandry

N = 120

Sr. No.	*Independent variables*	*Correlation coefficient ('r' value)*
I.	**Personal variables**	
1	Age	- 0.8888 *
2	Education	- 0.8411 *
II.	**Economical variables**	
1	Occupation	- 0.3014 *
2	Herd size	- 0.9464 *
3	Land holding	- 0.8881 *

contd...

III.	Social variables	
1	Family size	- 0.9322 *
2	Type of family	+ 0.8646 *
3	Number of children	- 0.8036 *
4	Age of marriage	- 0.8709 *

* Significant at 5 %

12.5.1 Age and Participation in Animal Husbandry

Data presented in Table 18 indicated that age of the respondent was observed negative significant with their participation in animal husbandry. It means that with decreasing in age of tribal farmwomen their participation in animal husbandry was increasing significantly. Here, the null hypothesis (H_0) was rejected.

7.5.2 Education and Participation in Animal Husbandry

Level of education of the farmwomen was negatively significant with their participation in animal husbandry. It shows that educated tribal farmwomen had poor interest in animal husbandry related activities such as breeding, feeding, care and management of animal are time consuming and laborious job. Hence, the null hypothesis (H_0) in case of education was rejected.

7.5.3 Occupation and Participation in Animal Husbandry

Data presented in Table 18 show that there was negative relationship between occupation of tribal farmwomen and their participation in animal husbandry. Hence the null hypothesis (H_0) in this case was rejected and it was concluded that there was significant negative relationship between occupation and their role in animal husbandry.

7.5.4 Herd Size and Participation in Animal Husbandry

The 'r' value presented in Table 18 indicates that the herd size possessed by tribal farmwomen and their participation in animal husbandry was observed negatively significant. It means with increasing the herd size the participation of tribal farmwomen was decreasing in animal husbandry. It might be due to the tribal farmwomen was engaged with multi dimensional activities like household, agriculture and animal husbandry. In other words it can be said that tribal farmwomen with small herd size had dominancy in playing role in animal husbandry. Here also the null hypothesis (H_0) was rejected in ease of herd size.

7.5.5 Land Holding and Participation in Animal Husbandry

The results presented in Table 18 show that there was negative significant relationship between land holding possessed by tribal farmwomen and participation in animal husbandry. It is obvious that the tribal farmwomen had big land holding was more engaged in various agricultural operations and less time avail for animal husbandry. It is concluded that the tribal farmwomen having big land holding was less involved in animal husbandry and hence, the null hypothesis (H_0) incase of land holding was rejected.

7.5.6 Family Size and Participation in Animal Husbandry

The 'r' value presented in Table 18 shows that there was negative significant relationship of family size with the participation of tribal farmwomen in animal husbandry. It was seen that the tribal farmwomen with large family size were more engaged in household activities and they devoted less time in animal husbandry. It is therefore null hypothesis (H_0) in this case was rejected.

7.5.7 Type of Family and Participation in Animal Husbandry

Data presented in Table 18 concluded that there was positive significant relationship between type of family of tribal farmwomen and their participation in animal husbandry. It means that the less participation of tribal farmwomen in animal husbandry was seen higher in case of joint family and less in case of nuclear family. Here, the null hypothesis (H_0) is rejected in case of type of family.

7.5.8 Number of Children and Participation in Animal Husbandry

Results revealed that there was negative significant relationship between the number of children and the participation of the tribal farmwomen in animal husbandry. It means the tribal farmwomen with less numbers of children, spared more time for animal husbandry. Hence the null hypothesis (H_0) incase of number of children was also rejected.

7.5.9 Age of Marriage and Participation in Animal Husbandry

Data presented in Table 18 shows that there was negative significant relationship between age at marriage and participation in animal husbandry. It means the tribal farmwomen got married at early age had more participation in animal husbandry as compared to tribal farmwomen got married at old age. Here, the null hypothesis (H_0) was rejected.

8. SUMMARY AND CONCLUSION

In this portion of the assignment the findings of the study have been given in a nutshell.

Major findings

The major findings are summarised below.

Socio-economic characteristics of the tribal farmwomen

1. **Age:** Slightly less than half the tribal farmwomen (48.33 per cent) were from middle age group.
2. **Education :** Half of the tribal farmwomen were found with primary level of education, followed by 39.16 per cent with no any formal education and only 10 per cent were educated upto high school and higher secondary level.
3. **Occupation :** A very large majority of the tribal farmwomen had (89.17 per cent) had household + farming alongwith animal husbandry as their main occupation.
4. **Herd size:** Majority (70.83 per cent) of the tribal farmwomen had small to medium size of herd.
5. **Land holding:** Major segments (34.17 per cent) of the tribal farmwomen had marginal size of land holding, followed by small (28.33 per cent) and medium (22.50 per cent) size of land holding.
6. **Number of children :** Majority (54.17 per cent) of the tribal farmwomen had 3 to 5 number of children followed by 25.83 per cent had 2 children and only 14.17 per cent of farmwomen had more than 5 children.

7. **Family size :** Slightly more than half (52.50 per cent) of the tribal farmwomen had medium size of family, followed by small (33.33 per cent) and only 14.17 per cent had large size of family.
8. **Type of family:** Majority (63.33 per cent) of the tribal farmwomen had nuclear type of family.
9. **Age of the marriage :** More than half (56.67 per cent) of the farm-women were married at the age of 18 years followed by 43.33 per cent married after the age of 18 years.
10. **Participation of tribal farmwomen in agriculture :** The highest participation of tribal farmwomen was seen in sowing/ transplanting followed by stubble collection, clod crushing, manuring and seedbed preparation/nursery management.
11. **Participation of tribal farmwomen in interculturing:** The highest participation of the tribal farmwomen was observed in weeding followed by gap filling, application of fertilizer, bird scaring, irrigation, banding and hoeing with hand, respectively.
12. **Participation of tribal farmwomen in harvesting and post-harvesting :** The highest participation of the respondents were in nipping/picking and threshing followed by harvesting, winnowing, storage, making threshing yard, bagging/packing and marketing of agricultural produce, respectively.
13. **Participation of tribal farmwomen in animal husbandry:** The highest mean score of participation of tribal farmwomen was obtained in case of cutting and bringing a fodder. Other participation of tribal farmwomen was seen in compost making followed by watering and feeding to animal, milking, cleaning of cattleshed, selling of milk and its products, bathing of animals, preparation of milk products, grazing of animals, taking animal for bull and veterinary services.

Involvement of the tribal farmwomen in decision-making

14. **Decision making about home management:** The tribal farmwomen were taking self decision incase of decoration taking of house (79.17 per cent) and selection and preparation of food (70.83 per cent). The husband dominated decision making events were house repair (79.17 per cent) followed by borrowing money for home management, repayment of

loan, manner of saving, repayment of loan, manner of saving, children's education, children's occupation and construction, children's occupation and construction of new house, respectively.

Joint decision was dominated in case of children's marriage (81.67 per cent) followed by selling and purchasing of ornaments, construction of new house and manner of saving, respectively.

15. **Decision making about farm management :** The majority of the decisions regarding farm management were not performed by tribal farmwomen, thus they were husband dominated decisions such as, when to irrigate the fields (87.50 per cent), quantity and type of fertilizers to be used in the farm (85.00 per cent) introduction of new crop variety, buying farm machinery/equipment, using plant protection measures, borrowing money for farm operation, installing oil engine, electric motor and pumps, selection of seed, deciding area to be sown under each crop etc.

    The joint decision was made only incase of hiring farm laborers (66.67 per cent), buying and selling of land and selling of surplus farm produce. It can be concluded that the tribal farmwomen did not play dominant role in making important decisions of farm management.

16. **Decision making about animal husbandry:** The decisions about selection of fodder and feed was dominated by tribal farmwomen (81.67 per cent) followed by sale of milk and its products. The jointly taken decisions were sale and purchase of animals (65.00 per cent) followed by selection of animal breed and keeping size of herd.

    It can be concluded that the important decisions regarding animal husbandry were dominated by tribal farmwomen themselves. The husbands remain recessive in decision-making regarding animal husbandry.

17. **Correlation study of crop husbandry:** The positive significant relationship was seen between the role of tribal farmwomen in crop husbandry and their age, education, herd size, land holding, family size and number of children. Whereas, the

negative relationship was observed among occupation, type of family, age at marriage and their role in crop husbandry.

18. **Correlation study of animal husbandry:** The relationship between all independent variables and the role of tribal farmwomen was observed negative except type of family.

Implication of the Study

Some important implication emerging from the findings of the study are presented below. During the study it was seen that many agricultural, animal husbandry related and house hold activities were performed by the tribal women, but when questions comes to take decisions regarding all these matter, their roles were seen meagre. As tribal farmwomen are the key units of the family of the tribals' community and they have skill in talking good decisions regarding all economic activities of the family, their participation in decision talking process needs to be encouraged by those people who are involved in the development of tribals.

Special measures need to be taken to increase the enrolments of girls in schools and to impart non-formal education for the dropouts, so that they would be able to keep accounts and made wage distribution to labours.

There is need to strengthen informal tribal education programme as means to the develop farm and home by providing modern agricultural technologies. The tribal farmwomen should be given training for another productive work outside the home especially for marginal and small size of holding. The family planning programme should be made more popular in this area.

The opportunities for productive employment should be enhanced by establishing agro-based industries, which might be organized in form of co-operatives. The tribal farmwomen should be allowed to take active part in decision-making process. These decisions will be more rational and practical oriented.

Suggestions for Further Study

1. Similar investigation may be conducted in other talukus of the district, so that the results of the study can be strengthened.
2. Similar studies may be conducted periodically with large sample to determine the role of tribal farmwomen in agriculture.

3. Similar studies may be conducted on the role of trained and untrained tribal farmwomen in agriculture and animal husbandry.

REFERENCES

1. Antoniades, A. and Papayiannis, C. (2000). The role of women in the family farm of the mountain region of Cyprus. Agricultural-Economics-Report, Cyprus Agricultural Research Institute, Nicosia, 39, pp. 12.
2. Badiger, C. and Rao, A. K. S. (1985). "Study on the participation of rural women in decision–making on farm and home aspect in Dharwad district of Karnataka state. J. of farming system, (3 & 4): 99.
3. Bagwe, A. N. (1991). Women caste : Case studies from Masure, India. Dissertation – Abstract International. 51 (9) : 3120.
4. Beck, M. M. and Swanson, J. C. (2003). Value-added animal agriculture: inclusion of race and gender in the professional formula. J. of Animal Science, 81 (11) : 2895-2903.
5. Bisht, B. S.; Bhuchar, S. K.; Pushpa, Pant; Kothyari, B. P.; Palni, L.M.S.; Pant, P; Allen, R. (ed.); Schreier, H. (ed.); Brown, S. (ed.); Shah, P. B. (2000). The pivotal role of women in the hills: gender analysis in Arah village in Uttar Pradesh, central Himalayas, India.

Proceedings of a workshop held in Baoshan, Yunnan Province, China, March 2-5 1999, pp. 35-46.

6. Chaudhary, M. D. and Ganorkar, P. L. (1993). Involving farm women in agricultural activities. Kurukshetra. 1 : 25.
7. Devdas, P. R.; Mathu, S. and Thangamant, K. (1972). Role of selected farm women in agricultural operation. Ind. J. Home Sci., 1: 50.
8. Divan, Yogita (2000). Major roles performed by the tribal farmwomen in maize production activities. Unpublished Ph. D. Thesis. A.A.U., Anand.
9. Everact, H. (1994). The professional life of a women farmer. Agricontact. 257 : 3-5.
10. Firdous, F.; Shazia, Y. and Nazia-Malik (2002). Role and status of women in Baluch families: A case study of rural community in District Faisalabad. Pakistan J. of Applied Sci., 2 (5) : 593-595.
11. Fremont, J. M. (2001). Agriculture in Europe. The spotlight on women. Statistics-in-Focus-Agriculture and Fisheries. 7 : 7-9.
12. Garcia, J. M. (1992). Women's work on Spanish family farms. Revista-de-Estudios-Agri-Sociales. 161 : 71-97.
13. Gasson, R. (1980). Role of farmwomen in England. Sociology Ruralis, 20 (3) : 165-180.
14. Geeta, K.; Chhaya, B.; Kalakannavar, G. and Badiger, C. (2000). Role performance, knowledge and opinion level of panchayat women members. Karnataka J. of Agril. Sci., 13 (1) : 130-133.
15. Habib, N. (1996). Invisible farmers in Pakistan : A study on the role or women in Agril. and impact of pesticides on them. pp. 129.
16. Hirannand and Kumar, K. (1980). Role of farmwomen of dry farming tract in decision making. Indian Co-operative Review, 17 (1) : 105.
17. Khanduri, B. K.; Rukma, R. and Rawat, R. (2004). Role of women in sustaining life in Garhwal Himalaya, Uttaranchal. Economic-Affairs-Calcutta. 48(3): 152-161.
18. Kulkarni, M. V. and Nandapurkar, G. G. (1991). Participation of rural women in decision making. Maha. J. Ext. Edu., 12 (2) : 307.
19. Kulkarni, M. V.; Nandapurkar, G. G. and Chitnis, D. H. (1990). Knowldege of Farm women regarding improved agricultural practices. Maha. J. Ext. Edu., 9 : 143.

20. Lepcha, T. (1987). A study of farmwomen's contribution to agricultural production in east district of Sikkim state. Unpublished M. Sc. (Agri.) Thesis, G.A.U., Anand.
21. Lohani, S.N. and Khatrichhetri, K. (2001). Women in sustainable agriculture development and environment. Ministry of Agriculture and Cooperatives, Agriculture and environment: communication issue, Kathmandu (Nepal). pp. 7-14.
22. Malkit Kuar and Sharma, M. L. (1985). "Role of rural women in animal husbandry : A study of Haryana". Indian Dairy Man, 37 (4): 20.
23. Mishra, K. (1991). Women in a tribal community : A study of Arunachal Pradesh. pp. 95.
24. Mrinali, G.; Bhowmick, B. C. and Gogoi, M. (2004). Factors determining the demand for female labour in farm activities. Economic-Affairs-Calcutta. 48 (3): 142-151.
25. Nasreen, R.; Qazi, A. R. and Ijaz, K. (1996). Role of women in decision making in various family affairs. J. Rural Dev. & Adm., 28 (3) : 85-90.
26. Nostrakis, A. M. (1992). The invisible variable dissertation. Abstracts International. 52 (9) : 3453.
27. Ogier, M. (1998). Possible valorization of the women's role in tropical agriculture. Unversite de Montpellier – II.
28. Parekh, P. M. (1990). Role expectation and role performance of rural women in farm management. A study in Valsad district of Gujarat state. Unpublished Ph. D. Thesis. G.A.U., Navsari.
29. Patel, A. J.; Patel, B. T.; Pandya, D. N. and Vihol, D. P. (1987). "Role of farmwomen in Agriculture". Unpublished paper. G.A.U., S. K. Nagar.
30. Qamar, Z.; Zahira, B.; Rehman, S. and Hammad, B. (2002). Role of skilled and unskilled factory working women in the rural economy of Punjab: A case study in Faisalabad. International J. of Agriculture and Biology, 4 (2) : 288-290.
31. Rathore, O. S. and Gaur, K. S. K. (1996). A comparative study on contribution and time utilization pattern of farm women in tribal and non-tribal areas. 80 (1) : 87-92.
32. Rattin, S. (1983). Choosing to be a farming women. Cahiers-de-statisique Agricole, 3-6 (35-37) : 11.

33. Rossier, R. C. (1992). Swiss women farmers. Their work on the farm. Schriflemreihe-der-Eidgenossischem. 250 (36) : 7.
34. Salick, J. (1992). Subsistence and the single women among the Amuesha of the Upper Amazon, Peru. Society and Natural Resources, 5 (1) : 37-51.
35. Saxena, D. and Bhatnagar, S. R. (1985). "A comparative study of time utilization pattern of tribal and non-tribal women in agriculture. Ind. J. Ext. Edu., 21 (1 & 2) : 83.
36. Sharma, K. D.; Pathania, M. S. and Vashist, G. D. (2003). Role of rural women in small scale agro-processing sector – an economic analysis of Samridhi Mahila processing co-operative society in Himachal Pradesh. Indian J. of Agril. Econ. 58 (3) : 577-588.
37. Singh, B. P. (1989). Need for effective utilization of women power resources in India. Economic affairs Calcutta, 34 (3) : 201-208.
38. Sinn, R., Ketzis, J.; Chem, T. and Haenilin, G. F. W. (1999). The role of women in the sheep and goat sector. Special issue. South Korea. 38 (8) : 259-569.
39. Sisodia, J. S. (1985). "Role of farmwomen in agriculture". Ind. J. Agril. Econ., 40 (3) : 228.
40. Thakor, R. F. and Patel. K. F. (1993). Factors related with farm women contribution in agriculture. Paper presented in the seminar on "Role of rural women in development" held at G.A.U., Anand on 8th Feb. 1992.
41. Uwakah, C. T.; Uwaeqbute, A. C.; Madukwe, M. C. and Doss, C. R. (1991). The role of women as farmers in eastern Nigeria.
42. Vagani, K. S. (1987). Patterns of involvement and agricultural women in household works and agricultural activities in Junagadh district of Gujarat state. Unpublished M. Sc. (Agri.) Thesis. G.A.U., Junagadh, Gujarat.
43. Vicente, J. (1989). Women in European agriculture. Agricultura-Y-Sociedud, 51 : 205-228.

Appendices

Interview schedule

Respondent's No. : ____________ Village : ____________

Date of interview : ____________ Taluku : ____________

PART – I

Personal and socio-economic characteristic

1. **Name :** ______________________________
2. **Age (years completed) :** ______________________________
3. **Education level :** ______________________________

   (a) Illiterate (b) Literate
4. **Occupation :**

   (a) Household + Farming + Agricultural labour

   (b) Household + Farming + Animal husbandry
5. **Herd size :**

   (a) Not owning an animal

   (b) Owning upto 2 animals

   (c) Owing 3 to 5 animals

   (d) Owing above 5 animals
6. **Land holding**

   (a) Marginal (up to 2.00 hectares)

   (b) Small (1.1 to 2.00 hectares)

   (c) Medium (2.1 to 4.00 hectares)

   (d) Large (Above 4.00 hectares)
7. **Family size**

   (a) Small family (up to 5 members)

   (b) Medium family (6 to 8 members)

   (c) Large family (above 8 members)
8. **Family type**

   (a) Nuclear type

   (b) Joint type

9. No of children

(a) No child

(b) Upto 2 children

(c) 3 to 5 children

(d) Above 5 children

10. Age at marriage

(a) Age at marriage 18 years of below

(b) Age at marriage above 18 years

PART – II

Frequency of participation of tribal farmwomen in agriculture and animal husbandry practices

(A) Farming operation

Sr. No.	*Farming operation*	*Participation*			
		Most frequent	*Frequent*	*Rarely*	*Never*
1	Ploughing/land preparation				
2	Clod crushing				
3	Stubble collection				
4	Manuring				
5	Preparation of seedbed				
6	Sowing/transplanting				
7	Banding after sowing				
8	Gap filling				
9	Weeding				
10	Hoeing with hand				
11	Application of fertilizers				
12	Spraying of insecticides				
13	Irrigation				
14	Bird searing				
15	Harvesting				

contd...

16	Making threshing yard				
17	Nipping/picking & threshing				
18	Bagging and packing				
19	Marketing				
20	Storage of farm produce				
21	Keeping accounts/giving wages to labours				

(b) Animal husbandry activities

Sr. No.	*Animal husbandry activities*	*Participation*			
		Most frequent	*Frequent*	*Rarely*	*Never*
1	Cutting and bringing fodder				
2	Watering and feeding to animal				
3	Bathing to cattle				
4	Shed cleaning				
5	Milking				
6	Preparation of milk products (Ghee, curd etc.)				
7	Compost making				
8	Selling of milk and its product				
9	Taking cows and buffalows for bull service				
10	Taking animals for veterinary services				
11	Grazing				
12	Any				

PART – III

Tribal farmwomen's involvement in decision-making process

(A) Home management

Sr. No.	*Home management*	*Participation*			
		Only self	*Husband*	*Alongwith husband/ family members*	*Not related*
1	Construction new house				
2	Decoration of house				
3	Children's education				
4	Children's occupation				
5	Children's marriage				
6	House repair				
7	Purchasing of household articles				
8	Selling and purchasing of ornaments				
9	Selection and preparation of food				
10	Borrowing money for home management				
11	Repayment of loan				
12	Manner of saving				

(b) Farm management

Sr. No.	Farm management	Participation			
		Only self	Husband	Alongwith husband/ family members	Not related
1	Deciding area to be sown under different crops				
2	Introduction of a new crop variety				
3	Selection of seed				
4	When to irrigate fields				
5	Quantity and type of fertilizers used on the farm				
6	Using plant protection measures				
7	Hiring farm labours				
8	Buying farm machinery / equipment				
9	Installing oil engine and electric motor				
10	Buying and selling of land				
11	Borrowing money for farm operations				
12	Selling of surplus farm produce				

(c) Animal husbandry

Sr. No.	*Animal husbandry*	*Participation*			
		Only self	*Husband*	*Alongwith husband/ family members*	*Not related*
1	Selection of animal breed				
2	Selection of fodder and feed				
3	Sale and purchase of animals				
4	Sale and milk and its products				
5	Keeping size of herd				